Make:
Volume 10
technology on your time™

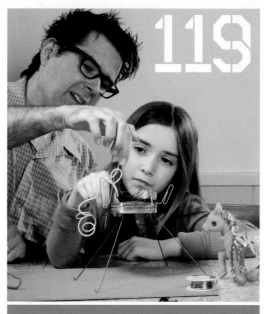

119

ON THE COVER: Nine-year-old Sarina Frauenfelder and her dad, Editor-in-Chief Mark Frauenfelder, use a continuity probe to test the wiring of the "Vibrobot," which uses an unbalanced toy motor to skim across the floor. Photographed for MAKE by Gregg Segal.

ROBOHOUSE How to set up your home to take care of itself and you.

72

Vol. 10, May 2007. MAKE (ISSN 1556-2336) is published quarterly by O'Reilly Media, Inc. in the months of March, May, August, and November. O'Reilly Media is located at 1005 Gravenstein Hwy. North, Sebastopol, CA 95472, (707) 827-7000. SUBSCRIPTIONS: Send all subscription requests to MAKE, P.O. Box 17046, North Hollywood, CA 91615-9588 or subscribe online at makezine.com/offer or via phone at (866) 289-8847 (U.S. and Canada); all other countries call (818) 487-2037. Subscriptions are available for $34.95 for 1 year (4 quarterly issues) in the United States; in Canada: $39.95 USD; all other countries: $49.95 USD. Periodicals Postage Paid at Sebastopol, CA, and at additional mailing offices. POSTMASTER: Send address changes to MAKE, P.O. Box 17046, North Hollywood, CA 91615-9588.

Make: Projects

The Brain Machine
Get altered states of consciousness with this microcontroller-driven sound and light device.
By Mitch Altman

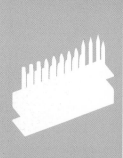
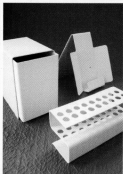

Plastic Fantastic Desk Set
Saw, drill, and bend your own objects made of ABS plastic.
By Charles Platt

Tabletop Biosphere
A fun demonstration of the ecological cycles that keep us alive.
By Martin John Brown

Electronic Test Equipment
See and understand what's happening inside a circuit. By Tom Anderson and Wendell Anderson

Maker Faire®

Build, craft, hack, play, MAKE.

BAY AREA: May 19 & 20, 2007
SAN MATEO FAIRGROUNDS

AUSTIN: October 20 & 21, 2007
TRAVIS COUNTY FAIRGROUNDS

MakerFaire.com

produced by **Make:**
makezine.com

sponsored by

Make:
technology on your time™

Volume 10

READ ME: Always check the URL associated with each project before you get started. There may be important updates or corrections.

Maker

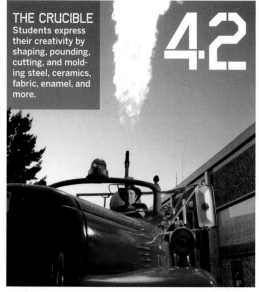

THE CRUCIBLE
Students express their creativity by shaping, pounding, cutting, and molding steel, ceramics, fabric, enamel, and more.

4.2

DIY

123

Make:
technology on your time™

EDITOR AND PUBLISHER
Dale Dougherty
dale@oreilly.com

EDITOR-IN-CHIEF
Mark Frauenfelder
markf@oreilly.com

CREATIVE DIRECTOR
Daniel Carter
dcarter@oreilly.com

MANAGING EDITOR
Shawn Connally
shawn@oreilly.com

DESIGNER
Katie Wilson

ASSOCIATE MANAGING EDITOR
Goli Mohammadi

PRODUCTION DESIGNER
Gerry Arrington

SENIOR EDITOR
Phillip Torrone
pt@makezine.com

PHOTO EDITOR
Sam Murphy
smurphy@oreilly.com

PROJECTS EDITOR
Paul Spinrad
pspinrad@makezine.com

ASSOCIATE PUBLISHER
Dan Woods
dan@oreilly.com

STAFF EDITOR
Arwen O'Reilly

CIRCULATION DIRECTOR
Heather Harmon

COPY CHIEF
Keith Hammond

SALES & MARKETING ASSOCIATE
Katie Dougherty

EDITOR AT LARGE
David Pescovitz

MARKETING & EVENTS COORDINATOR
Rob Bullington

VIDEO PRODUCER
Bre Pettis

ONLINE MANAGER
Terrie Miller

MAKE TECHNICAL ADVISORY BOARD
**Gareth Branwyn, Limor Fried, Joe Grand,
Saul Griffith, William Gurstelle, Bunnie Huang,
Tom Igoe, Mister Jalopy, Steve Lodefink, Erica Sadun**

PUBLISHED BY O'REILLY MEDIA, INC.
Tim O'Reilly, CEO
Laura Baldwin, COO

Visit us online at makezine.com
Comments may be sent to editor@makezine.com

For advertising and sponsorship inquiries, contact:
Katie Dougherty, 707-827-7272, katie@oreilly.com

Customer Service cs@readerservices.makezine.com
Manage your account online, including change of address at:
makezine.com/account
866-289-8847 toll-free in U.S. and Canada
818-487-2037, 5a.m.–5p.m., PST

MAKE is printed on recycled paper with 10% post-consumer waste and is acid-free. Subscriber copies of MAKE, Volume 10, were shipped in recyclable plastic bags.

PLEASE NOTE: Technology, the laws, and limitations imposed by manufacturers and content owners are constantly changing. Thus, some of the projects described may not work, may be inconsistent with current laws or user agreements, or may damage or adversely affect some equipment.
Your safety is your own responsibility, including proper use of equipment and safety gear, and determining whether you have adequate skill and experience. Power tools, electricity, and other resources used for these projects are dangerous, unless used properly and with adequate precautions, including safety gear. Some illustrative photos do not depict safety precautions or equipment, in order to show the project steps more clearly. These projects are not intended for use by children.
Use of the instructions and suggestions in MAKE is at your own risk. O'Reilly Media, Inc., disclaims all responsibility for any resulting damage, injury, or expense. It is your responsibility to make sure that your activities comply with applicable laws, including copyright.

Contributing Editors: William Gurstelle, Mister Jalopy, Brian Jepson

Contributing Artists: Roy Doty, Steve Double, Nick Dragotta, Amber Henshaw, Dustin Amery Hostetler, Timmy Kucynda, Tim Lillis, Charles Platt, Greg Ruffing, Nik Schulz, Damien Scogin, Gregg Segal, Robyn Twomey

Contributing Writers: Mitch Altman, Tim Anderson, Tom Anderson, Wendell Anderson, David Battino, Kipp Bradford, Gareth Branwyn, Martin John Brown, Matt Coohill, Ken Delahoussaye, Cory Doctorow, George Dyson, Ken Gracey, Joe Grand, Saul Griffith, Amber Henshaw, Tom Igoe, Tom Jennings, Brian Jepson, John Krewson, Tod E. Kurt, Todd Lappin, Steve Lodefink, Stephanie Maksylewich, Brian Nadel, Tim O'Reilly, Ross Orr, Tom Owad, John Edgar Park, Tom Parker, David Pescovitz, Bre Pettis, Charles Platt, Michael H. Pryor, Douglas Repetto, McKinley Rodriguez, George Sanger, Donald Simanek, Dave Sims, Matthew Sparkes, Bruce Sterling, Andrea Steves, Nathan True, Andrew Turner, Cy Tymony, Jason Verlinde, Michael Wernecke, Megan Mansell Williams, Tom Zimmerman, Lee D. Zlotoff

Interns: Matthew Dalton (engr.), Adrienne Foreman (web), Jake McKenzie (engr.)

Contributors

Mitch Altman (*Brain Machine*) was a geeky kid who dreamed about electronics when he was 3. He's been trying to find worthwhile things to do with this skill ever since. He helps the world enjoy turning TVs off with his invention TV-B-Gone. He's also started a nonprofit vegetarian restaurant, started a rural commune, and co-founded 3ware, a Silicon Valley company. Altman still has time to meditate, do cool volunteer work, and hang with friends, and is currently developing a Sound & Light Machine that will help people sleep better.

Matthew A. Dalton (MAKE engineering intern) grew up in California, moving from place to place with his mom and little sister. He quickly learned how to fix whatever broke, and kept everything running smoothly. With an inclination toward electronics, Dalton dove into engineering at the local junior college, and can now be found struggling with homework on any given night. He's hopeful that seven years in the military fixing jets and ECM pods will help him in his endeavors. With several projects on the table, Dalton is just waiting for summer to hit. If only school didn't take up so much time!

Martin John Brown (*Tabletop Biosphere*) has written scientific papers about ultraviolet radiation and previously studied forest canopies. He's written for *Sierra*, *Air & Space/ Smithsonian*, and *E/The Environmental Magazine*, and is currently co-authoring a book about the social experience of knitting. *Knitalong* comes out in 2008 from Stewart, Tabori & Chang. More recently he's been bogged down in other projects, most notably building a tiny house, eating too much barbecued chicken, thinking a lot about the problem of dishwashing, and playing flamenco guitar rather badly. martinjohnbrown.net

Tom Parker (*Make Money*) is an author who lives in Ithaca, N.Y., and works for Cornell University. He has written and illustrated *Rules of Thumb* volumes 1 and 2, *In One Day*, *Never Trust a Calm Dog*, and *Le Livre du Bon Sens*, a French translation that even he can't understand. Parker and his books have appeared on *Late Night with David Letterman*, *The Tonight Show*, *The Today Show* and the *CBS Evening News,* but he notes that signed copies can be bought off eBay for pennies on the dollar. When he is not tinkering with junk, Parker is a flight instructor and flies a 1956 Cessna 180 bush plane.

Gregg Segal (Cover and MakeShift photography) studied photography and film at the California Institute of the Arts and went on to do a master's degree in dramatic writing at NYU. His penchant for drama now finds an outlet in photography. After a brief stint as a writer, he returned to shooting and has been documenting Los Angeles and its people for the past 12 years. "My mother tells me she knew I'd become a photographer when, after she got me a camera for my 11th birthday, I photographed our neighbor's garbage," he says. When not going for the jugular in his personal work, Segal shoots for *Esquire*, *Premiere*, *Dwell*, *Newsweek*, and *Entertainment Weekly*, among others.

When he's not talking to garage ski builders, Seattle resident **Jason Verlinde** (*Downhill Makers*) can usually be found interviewing bands. Verlinde is the publisher of the *Fretboard Journal* (fretboardjournal.com), a quarterly magazine documenting musician culture and guitar builders. Now that the weather has cleared up in the Northwest, there's a good chance you'll find him working on his old Toyota Land Cruiser in his driveway or walking his dogs.

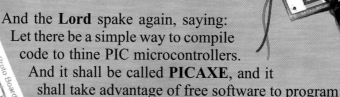

WE'RE ALL ALRIGHT

By Shawn Connally

RECENTLY, I HEARD THE SONG "ALRIGHT GUY" ON MY WAY TO WORK.

I think I'm an alright guy
I just wanna live until I gotta die
I know I ain't perfect but God knows I try
I think I'm an alright guy
I think I'm alright.

Singer/songwriter Todd Snider goes on to say that while he might occasionally get drunk and even a little rowdy, he doesn't have a trunk full of dead bodies.

Maybe it was the lack of coffee in my system, but this struck me as quite a profound statement. Something patently obvious but culturally ignored. We're all pretty OK, us ordinary folk. We're all alright. We have jobs, hang out with friends and family, take up hobbies, and make some mistakes along the way.

All the while, media and popular culture tend to put us down. TV shows make fun of our mistakes and imperfections, even our passions, however misplaced they may be. There's always someone on a TV sitcom more than happy to make fun of the normal guy. And reality shows like *American Idol* excel at making fun of people singing their guts out — the more off-key, the better for ridiculing!

As a society we're buying into it, putdown by putdown. You're too short, too fat, too old, too poor, too young, too klutzy, too dumb, too imperfect.

But then I realized, hey, this is where MAKE is different. Perhaps that's the secret of our success. We give our readers instructions to build things themselves, we celebrate backyard hacks and garage-grown innovations, and we think *everyone* is smart enough to understand what we're writing about. We assume the best in people, not the worst.

Mainstream culture tends not only to highlight our shortcomings, but also to celebrate only celebrities — the biggest, brightest, richest, wildest, prettiest, most handsome.

On the flipside, MAKE celebrates ordinary, alright folk. We're talking about homemade BBQ grills stuffed into newspaper boxes, and espresso machines made from mortar shells (*pages 20-21*). Perhaps neither is the perfect machine for the job, but they're imaginative and handmade by ordinary people. And that's why we like them.

We're hoping this volume adds to that home-grown community and spirit. I suggest you start out with the uplifting "Fail Early! Fail Often!" (*page 34*), which explains the merits of failure, why it's not a dirty word, and how failing is always farther along the road to success than not trying at all.

Our focus is on Home Electronics this issue, with a special section full of great projects and articles, from top Roomba hacks (*page 67*) to making your house into the ultimate robot (*page 72*). The Primer on electronic test equipment (*page 158*) includes a refresher on the basics of voltage, current, and resistance. It's one article I'll keep at my home electronics workbench (*page 58*) for quick reference.

We'll show you how to literally build your own world with our Biosphere project (*page 110*), and how to make your own brain machine to see how lights and sounds affect the mind (*page 88*).

You'll read in amazement about high-power rock-etry amateurs (*page 48*), self-replicating machines (*page 38*), and the fiery world of The Crucible (*page 42*). These are communities working on, failing at, and ultimately succeeding with their dreams.

Even our Reader Input (*page 176*) assures me that MAKE is focusing on the right sorts of people, and the right sorts of accomplishments. Enjoy 13-year-old Vinnie Brubaker's personal statement, "How MAKE Magazine Changed My Life," and the fantastic artwork and story created by third-graders in Oregon for our MakeShift contest.

If these types of stories, ponderings, and projects inspire you, then my advice is to surround yourself with them. Seek out ordinary people creating extra-ordinary projects, then revel in appreciating their work, their vision, and of course, their imperfections.

Shawn Connally is managing editor of MAKE.

▶▶ You know you're a Maker when:

You know how to solder and sew (or you aspire to)

You subscribe to MAKE & CRAFT (or you aspire to)

Sign up today to receive a free issue.

Enter promo code SUMMER

makezine.com/subscribe

craftzine.com/subscribe

News From the Future

STUFF LIKE SOFTWARE

By Tim O'Reilly

ONE OF THE THINGS WE LOOK FOR AT O'Reilly Media is people who know how to do magic. These are people who can surprise us by their mastery of technology, some trick or two that makes us start with surprise and say, "How did you do that?" As in the famous Arthur C. Clarke dictum, it isn't really magic, just something they know that we don't. Sometimes the magic is a simple hack or tool. Sometimes it's a really big, world-changing idea that they've just caught on to before everyone else.

The first websites were like that. In 1993, people's eyes bugged out when they saw someone click a link on what appeared to be a document on a PC screen, and suddenly pull up another document from a server halfway around the world. Now it's taken for granted; the magic has moved on. Today, the greatest magic appears to be in the area of manufacturing. Some people know how to make stuff, and others merely marvel.

Now I'm not just talking about the kind of making celebrated in these pages, the magic exposed, step by step. Cool as that is, there's nothing hidden. I'm talking about the kind of magic that MAKE columnist and kite enthusiast Saul Griffith hinted at in a recent conversation: "We sent in our new kite design on Monday and were testing it on Friday."

What's magical about that? It's a signal about the coming revolution in the manufacturing supply chain. You see, Saul's kites are designed on a custom CAD program, and prototypes are manufactured quantity one by a factory in China, and delivered halfway around the world in a matter of days.

I found myself marveling at a similar bit of magic when talking with Saul's colleague at Squid Labs, Colin Bulthaup, who has a new company called Potenco, commercializing the portable power generator they designed for the One Laptop per Child project. Colin casually showed me a table full of prototypes of the pull-string charger, all looking like they'd come fresh from a blister pack at Best Buy. How'd he do that? No handmade prototypes these.

It used to be hard to get stuff made. You made something yourself, or it was mass-produced. There wasn't a whole lot in between. But now, the internet and what Thomas Friedman calls "the flat world" are making it easier and cheaper to have not just a workshop but a factory at your beck and call.

"We're now able to iterate on a hardware design on a weekly basis," Colin says. "This is leading to rapid exploration of ideas in hardware analogous to what we've gotten used to in software. Up to a year ago, what you'd get were mockups. Now you can get fully functional mechanical prototypes." It's still relatively expensive to get these prototypes made, and there's more magic required to scale up to consumer (or mass customization) quantities, but the writing is already on the wall.

This news from the future affects not just makers who are starting to think about becoming entrepreneurs, but also consumers. As factories and supply chains become smarter, we're seeing a future of mass customization. And entrepreneurs are harnessing all the power of the internet to approach manufacturing in new, creative ways.

Threadless.com is a great example. Using Digg-like voting systems, they encourage users to submit, vote on, and discuss T-shirt designs, and then they manufacture and sell the most popular designs. Threadless was reportedly up to $20 million in revenue last year. If you have to get to scale before you can manufacture something, you can now aggregate the demand before spending a nickel.

All of this is not to mention that personal fabrication devices are following the same price curve that we saw with typesetting in the 1980s, leading up to the desktop publishing revolution. We've seen offers for a $2,000 laser cutter, down from $16,000 a year ago.

In short, get ready for a future in which stuff looks a lot more like software.

Check makezine.com/10/nff for related stories.

Tim O'Reilly (tim.oreilly.com) is founder and CEO of O'Reilly Media, Inc. See what's on the O'Reilly Radar at radar.oreilly.com.

THE CRUCIBLE

has classes that set your imagination on fire!

WELDIN
BLACKSMITHIN
FOUNDRY & MOLDMAKIN
NEON & LIGH
JEWELRY & ENAMELIN
FIRE PERFORMANC
KINETI
MACHINE SHC
WOODWORKIN
and mor

On-going classes
on weeknights and
weekends.

Visit
www.thecrucible.org
for class descriptions ar
schedule.

SPECIAL SUMMER PROGRAMMING

Fire Arts Festival
Weekend Classes – July 13 & 14

**Four-day workshops in blacksmithing, foundry
fundamentals, jewelry making, neon sculpture, and more**
August 20-23

Camps & Workshops for Young Makers (ages 8 to 18)
June 18-22, August 6-10, July 10 & 11
Orientation for interested families — May 5

Don't miss The Crucible's Seventh Annual

Fire Arts Festival

July 11 – 14
This spectacular open-air exhibition of fire performance, fire sculptur
and interactive fire art includes The Fire Odyssey, a theatrical
production that blends fire performance with ballet, opera, aerial
dance, and other performance styles.

Tickets available online at **www.thecrucible.org**

THE CRUCIBLE
260 7th Street Oakland, CA 94607 • 510.444.0919

Make Free

AGREE TO DISAGREE

By Cory Doctorow

IN MY LAST COLUMN, I WROTE ABOUT HOW we're being hemmed in by fake "agreements" that we make simply by moving through time and space. Chances are, the last box you opened has some kind of shrinkwrap license that makes you promise not to take apart whatever was inside of it. These sneaky, self-replicating pieces of legal code are a stealth attack on the maker spirit, and I've had enough of it. So I've got a plan — read on for more.

The content of these "agreements" is just as crazy as their form. It's typical to sign away your rights to reverse-engineer your property, to resell it, or to sue anyone if it blows your head off. An early version of the Windows Vista end-user license agreement (EULA) had you "agreeing" not to publish benchmarks on your system's performance. Amazon Unbox's EULA makes you promise to allow spyware on your computer. Receipts from Circuit City and Best Buy have an extra foot of paper tape telling you all the stuff you agreed to by being dumb enough to shop there.

Somehow, we've gone from a sign over the register that says, "All sales final," or even "Tipping is appreciated" and "No smoking, please," to "By entering these premises, you agree to let us come over to your house and punch your grandmother. This agreement is subject to change without notice."

As if. As if by opening a package, you can waive your rights. What if you're under 12? Drunk? Mentally unfit?

Enough is enough. I've just launched the Fair Agreement campaign — a campaign to ridicule and vilify these idiotic policies. Fair Agreement is built around the anti-EULA, first suggested by Boing Boing reader Steve Sitmus. The idea is to salt all your interactions with companies with the text shown at right.

Put the Fair Agreement on your utility bills. On the credit card slip at Best Buy. Paste it on the warranty cards that come with your next computer or video game. You could be the only person in the world who uses iTunes and is "lawfully" permitted to crack the DRM (but don't hog it — go ahead and get your friends to send in registration cards, too).

The Fair Agreement is ridiculous, but it's supposed to be ridiculous. That's the whole point — agreements

READ CAREFULLY. By [accepting this material/ accepting this payment/accepting this business card/ viewing this T-shirt/reading this sticker] you agree, on behalf of your employer, to release me from all obligations and waivers arising from any and all NON-NEGOTIATED agreements, licenses, terms of service, shrinkwrap, clickwrap, browsewrap, confidentiality, non-disclosure, non-compete, and acceptable use policies ("BOGUS AGREEMENTS") that I have entered into with your employer, its partners, licensors, agents, and assigns, in perpetuity, without prejudice to my ongoing rights and privileges. You further represent that you have the authority to release me from any BOGUS AGREEMENTS on behalf of your employer.

are noble beasts, arrived at through diplomacy and negotiation.

I've made it my sig file, and I'm making stickers and T-shirts available at cost on my website, reasonableagreement.org. It's all public domain, too — make your own versions, sell 'em or give 'em away. Just get the message out there.

You don't have any choice in the matter, anyway. Just by reading this column, you've agreed.

Cory Doctorow (craphound.com) is a science fiction novelist, blogger, and technology activist. He is co-editor of the popular weblog Boing Boing (boingboing.net), and a contributor to *Wired*, *Popular Science*, and *The New York Times*.

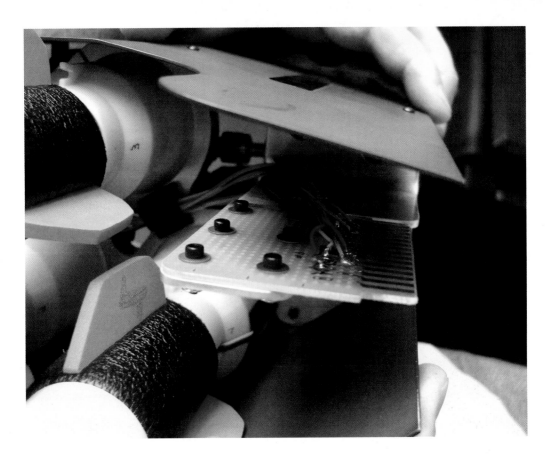

Hardcore Laser Tag

Mike Yates plays laser tag. But we're not talking about the kind of game you might find down at the Family Fun Center or around your neighborhood using some toy.

Yates is at the center of a group of hardcore laser tag players who prefer to play their games in warrens of bunkers below abandoned military bases, using highly modified custom taggers.

From his home base in Washington state he's been hacking laser tag gear for a decade or so, and his latest project, the Scorpion, is a custom sniper rifle and RPG launcher based on Hasbro's newest-generation Lazer Tag Team Ops (LTTO) platform.

Since Yates moves in the shadowy world of underground tagger developers, he was able to obtain a pre-release prototype circuit board for the Hasbro LTTO tagger, called the Tag Master Blaster, from its developers, Shoot the Moon Productions. The TMB is a laser tagger that also fires a foam-tipped, IR-emitting RPG round. To fire the rocket, the player hand-pumps the gun to pressurize a plastic pneumatic tank, which propels the rocket

20–40 feet. Very cool, but not very Yates.

For ten months, Yates and a fellow laser hacker built the ultimate LTTO-compatible IR weapons system. The Scorpion is mil-spec from top to bottom. The most obvious feature is the four-rocket magazine that attaches to the top of the weapon.

Four Mac pneumatic solenoid valves control the airflow from the 3,000-psi aluminum tank to the rockets through stainless steel plumbing. The trigger grip for firing the rockets comes from an actual tank and was found on eBay. In addition to being a rocket launcher, the weapon functions as a high-powered LTTO sniper rifle with a 1,500-foot range, thanks to the big 4-inch adjustable-focal-length lens.

There's a big element of one-upsmanship motivating the development of these guns. Everybody knows that when they show up at the annual NW TagFest, there will be some new showstopper there. More often than not, Mike Yates is behind it.

—*Steve Lodefink*

>> **Mike Yates' Custom Tag:** mysite.verizon.net/resobodw

Sound Effects

"Music comforts me," says sound sculpturist **Jeroen Diepenmaat** of the Netherlands. "By introducing music and sound in my pieces, I can give other people that comfort."

A self-professed vinyl addict, Diepenmaat's favorite sound-makers are records. In his work entitled *Pour des dents d'un blanc éclatant et saines*, he rigs a record player so that the bill of a stuffed bird, rather than a needle, releases the sound from an old LP.

In his Deventer home studio, which is subsidized by the Netherlands Foundation for Visual Arts, Design and Architecture, Diepenmaat creates sculptures born of materials gathered from secondhand shops and ideas picked up in unusual science books.

He also relies heavily on coincidence. Case in point: *Pour des dents*, shown recently at Gigantic ArtSpace in New York City, was inspired by a stuffed bird that his taxidermist father made for a friend.

Diepenmaat looked closely at the bird bill and realized that it could easily play a record if positioned just right. His dad stuffed a peewit for the job, and Diepenmaat wired the bird's feet to a thrift-store record player. Under the peewit's bill, he spun a record from his vast collection of birdsong LPs (yes, entire albums of twitters and trills are available).

While the record player's arm and needle remain eerily locked, the bird bill releases a scratchy, distant chirping from the vinyl's grooves. "I like the image of a coincidental character," Diepenmaat says. "Like the bird just flew in and was attracted to the spinning vinyl, and put his bill on it."

Of Diepenmaat's other sound sculptures, *Loop/loop* transforms records into wheels, while *The devil wears those hypocrite shoes* wires a pair of speaker-sneakers so that when the wearer walks, the shoes make noise. These days, the artist is working on a mechanical orchestra activated by bicycles.

—*Megan Mansell Williams*

≫ **Sound Sculptor:** jeroendiepenmaat.nl

Photograph by Jeroen Diepenmaat

Photograph by Josch Hambsch

Star Trails

Looking far, far into the past has been a passion for **Josch Hambsch** for more than 30 years, but it's taken the digital revolution to let him capture the mesmerizing trails of the stars as the Earth rotates.

An amateur astronomer for nearly half his life, Hambsch waxes poetic about the night sky and the capabilities of digital photography. "I can show things never observed with the naked eye," he says. "It's also a very special feeling when I am imaging objects that are millions or even billions of light years away. One can see very far into our past."

Explaining the technical side of his ethereal images, Hambsch, a nuclear scientist in Belgium, says the star trails images are actually 128 photos taken over an 11-hour period and then stitched together using a freeware program, aptly named StarTrails (startrails.de). "Each image was exposed for five minutes at 800 ASA. Each image shows a short movement of the stars across the sky due to the Earth's rotation, and only when all the images are combined is the nearly half-circle movement visible."

The digital revolution has done wonders for his work, both artistically and scientifically. He began doing star trails photography about ten years ago, but it wasn't until two years ago, when he used a DSLR digital camera, that he considered his work a success. He also enjoys using special, cooled CCD (charge-coupled device) cameras for imaging DSOs (deep sky objects). For those, he uses a medium-sized telescope, but for the star trails images all he needs is a wide-angle lens and a long exposure time.

Light pollution in his backyard is one of the reasons Hambsch travels to very dark places like Namibia to take his exceptional images of the night sky. But he says you can still get similar images in the more light-polluted Northern Hemisphere. It's just a matter of shorter exposure times and specialized software.

"If one thinks about the huge distances and the possibilities of the available technology to catch light that has traveled for up to several billion years, just to catch the small detector I am putting in its way, that is just fantastic." —*Shawn Connally*

≫ **More Photos:** www.astronomie.be/hambsch

Mortar Shells to Coffee Grounds

The area around the house of **Azmeraw Zeleke** in northern Ethiopia is littered with burnt-out mortar shells left over from a war with neighboring Eritrea.

For months, Azmeraw wondered what he could do with them as he saw them being sold around Mekele town (about 800km from the capital, Addis Ababa). They were being used for washing clothes or for crushing things. Finally, he struck upon the idea of converting the shells into the inner workings of coffee machines.

The shells stand about 1 meter high. Azmeraw cuts off the pointed ends, seals them, and puts holes in the aluminum cylinder. The cylinder then channels the water, coffee, and milk.

Coffee is a major export from Ethiopia and plays a big role in life. After meals, the traditional coffee ceremony allows family and friends to get together to share news and discuss the issues of the day. Coffee shops are also popular. Each of Azmeraw's machines costs about $1,300, which is relatively cheap compared to imported machines. A local coffee shop owner, Haile Abraha, says the machines work well and make great coffee.

Azmeraw thinks he has sold hundreds — he's not sure exactly how many — since he started production five or six years ago. But he says it can be difficult to convince people in the area to buy the machine because of the mortar shell. "These shells have all been used. We all need peace and we don't want war, but once these shells have been used, we should use our skills to do something with them," he says.

"Sometimes I think about the fact they were used for war, but I want to change them to do something good. They could be a symbol of war, but I am doing something good out of the bad."

Azmeraw has big plans for his small business. At the moment, he works out of three ramshackle rooms with gaps in the corrugated roof. His staff of six sells the machines to coffee shops and restaurants in the area. In the future, he hopes to sell them even farther afield — perhaps even to Eritrea.

—*Amber Henshaw*

Photography by Amber Henshaw

Hot Copy

Cookouts and cars go pretty well together, a fact that hasn't escaped custom grill maker **Steve Barker**. The former kitchen and bath renovator extended his love of muscle cars beyond souping up '55 Chevys in his spare time. He made BBQ history a couple of years ago when he tricked out the grill in his backyard to resemble a hot rod engine, attaching exhaust pipes to release smoke. Now, he's moved on to more pressing matters — a newspaper dispenser that serves steak instead of headlines.

The muscle car grill became such a conversation piece that Barker decided to build a BBQ business. Today, his cookers start at $1,300. In addition to the headers, he also welds diamond-plate shelves onto homemade stainless steel grill boxes, and uses Wysco pistons as control knobs. A shifter handle opens the lid, and each grill is mounted on an engine stand so that it swivels. Barker even customizes the grilling surface so racing fans can brand the name of their favorite engine (a Hemi, say) onto a strip steak.

The clever grills eventually caught the attention of a newspaper publisher in North Carolina. Instead of a hot rod, the client craved a grill made from an old newspaper box. Barker gutted, sandblasted, and powder coated a dilapidated vending machine, then installed a burner, an igniter, and a porcelain-coated grill. He also cut out the back to fit a propane tank and set the thing on wheels. "It came out perfectly," says Barker. "I even cooked a couple of hot dogs on it before I sent it out, just to test it."

But Barker's pièce de résistance might just be a BBQ he's assembling in the form of a baby grand piano. Five feet wide and 6 feet long, the lid opens into a dining table. Inside is a storage area, a refrigerator, even an icemaker. When he finishes the masterpiece, Barker intends to leave his native Ohio for Florida ... where you can grill all year long.

—*Megan Mansell Williams*

>> **Steve Barker's Grills:** musclecarbbqgrills.com

Eight-Step Program

What if buttons could talk back? **Monome**, a Philadelphia collective of musicians and tech-savvy designers, is exploring that concept in a big way. Its 40h "computer-human user interface" is a 6.75-inch-square USB controller on which each of the 64 backlit buttons is also a pixel in an interactive display. (40h is 64 in hexadecimal.)

Software for Windows, Mac, and Linux transforms the 40h into eight step faders, a one-bit video display, a drum machine, a spectrograph, an audio-sample slicer, triggers for popular music programs, or whatever you decide to make. Both the software and firmware are open source, and Monome purposely included extra inputs on the circuit boards to facilitate modding. Other makers have already added knobs, accelerometers, and even alternative LEDs.

"By all means we want you to open up and change your hardware," begins one of the detailed tutorials on the Monome site. "We are not responsible if something goes wrong, [but] we will help you fix your problem and we'll even repair your unit" (fee applies for that).

In reality, the 40h is more of an interactive art experiment than a retail product. Only 400 units were made, and custom parts and hand assembly drove the price to $500. Monome's manufacturing philosophy is meticulous, too. Following the Restriction of Hazardous Substances Directive (see wikipedia.org/wiki/ROHS), the group avoids lead-based components and other toxics. It seeks out local suppliers, uses recyclable packaging, and even delivers orders to the UPS depot by bicycle.

The name Monome (pronounced *mon ohm*) is a nod to the simplicity of monomial numbers, to the minimalist credo of doing more with less.

Addressing two of the most frequent user requests, Monome's Brian Crabtree writes, "If we had tri-colored LEDs and velocity-sensitive pads, it'd be an entirely different instrument. I am honestly less interested in gradients (tri-color or velocity) than I am in the 16×16 [an upcoming 256-button controller]. More bits!"

—*David Battino*

≫ **Monome:** monome.org

Photograph courtesy of *monome.org*

Rolling Mountain Thunder

Frank Van Zant wanted to make a statement that would last, and few things last as well as concrete in the desert. So when Van Zant's 1948 Chevy truck broke down two hours east of Reno, sometime in the summer of 1968, he decided to take it as a sign.

He set up camp and started scavenging — metal, glass, old cars, a typewriter — whatever he could get. Then he bought sacks of concrete, and without any art background, this retired sheriff and World War II vet began shaping Thunder Mountain Monument alongside Interstate 80, a memorial to the plight of Native Americans.

"No other person has duplicated what he did with concrete freehand," says Van Zant's son, Daniel Van Zant, who now owns the land and the structure outside of Imlay, Nev. "Concrete wants to drop. It's not easy to get that wet and heavy material to stay in place."

Daniel says his dad built the main monument building with 2×4 framing, marine-quality plywood, and concrete that he smeared on with his bare hands. "I never saw him use tools with the concrete.

His hands were always bloody and calloused, and sometimes he had to wait for them to heal up before he could go on."

Today, the structure sits open and unattended, a roadside attraction that evokes a different time. Back in the day, young hippies on the road west would stop and accept shelter in a hostel building on the site, helping out and learning from the sculptor who by then had taken the name Chief Rolling Mountain Thunder.

But the hostel eventually burned down, the 60s became the 70s and then the 80s, and Van Zant ended his days alone at the monument — just him and the 200 statues of Indians he had fashioned with wet concrete and his bare hands.

—Dave Sims

≫ **Thunder Mountain:** thundermountainmonument.com

Photograph by Richard Menzies

THE REAL RENAISSANCE MAN

BENVENUTO CELLINI REJECTED HUMAN LIMITATIONS TO PROVE HIS VALOR.

By Bruce Sterling

Benvenuto CELLINI, sculpteur et orfèvre
né en 1500, mort en 1571 à Florence
Œuvres : Nymphe de Fontainebleau, Persée,
Jupiter Tonnant, etc.

A GOLDSMITH NEEDS TO LEARN FORGING, casting, filing, soldering, piercing, sawing, and polishing. It's a difficult craft to master. Now suppose this same goldsmith was also a fine musician. And a painter. Suppose that he also carved large statues in marble, and cast them in bronze. Let's further suppose that he was a swordsman, a crack shot, a master of artillery, an occultist, and a part-time cleric, and that he personally killed uncounted numbers of people. What kind of guy would that be?

He'd have to be a "Renaissance man." Nowadays, many tinkerers, hackers, hobbyists, and dilettantes fancy themselves to be Renaissance men. Benvenuto Cellini's autobiography offers unique personal insight into how it felt to be a genuine Renaissance man in the genuine Renaissance. Because Benvenuto Cellini was that guy; he was there, on the ground, doing it.

Cellini never mentions the word *Renaissance*. He doesn't know he's having one. Every once in a while, some pope or duke will vaguely refer to the glorious rebirth of ancient learning. Still, Cellini's world is, by our standards, a bloody catastrophe. The graveyards brim over with disease. Looted cities are aflame. Traitors, bandits, assassins, and poisoners lurk around every street corner.

Narrating his career from the perspective of his mid-50s, Cellini is frankly astounded that he has survived so long. He sees his own lifetime — the glorious peak of the Italian High Renaissance — as one long calamity.

To judge by his own words, Cellini himself is a violent megalomaniac. Mind you, Cellini was definitely a great Renaissance artist: after all, Cellini handcrafted the awesome golden *Saliera*, unquestionably the greatest, grandest, goldest salt-and-pepper shaker ever made. But he's also unbelievably touchy, and generally heavily armed. The guy is chop-lickingly fond of every form of weaponry: he's got dirks, daggers, poniards, swords, and big two-handed broadswords, plus pikes, fowling guns, and arquebuses. He really comes into his own when he's given command of an artillery squadron.

Perseus is straight out of an R-rated splatter film, but it was a major hit with the Florentines from the moment Cellini unveiled it in 1554.

Cellini's famous statue, *Perseus and Medusa*, is his proudest achievement. We've all seen pics of this statue; it's so wreathed in solemn art history that it's hard for us moderns to understand what we're seeing. The staue portrays a stark-naked young man trampling the nude corpse of a decapitated Medusa while he waggles her severed head in a dripping welter of gore. *Perseus* is straight out of an R-rated splatter film, but it was a major hit with the Florentines from the moment Cellini unveiled it. They were so thrilled they wrote sonnets about it.

In the Italian Renaissance, staying alive was for sissies. Cellini fought in the gruesome Sack of Rome in 1527: overwhelmed by a vast armed mob of looters and arsonists, the city's population was cut in half. When he traveled home to Florence to see his beloved dad, he found his own family wiped out by the Black Death, and the family house occupied by a demented, screaming hunchback. These calamities were happening to an internationally

Photography by Corbis

CELLINI'S FLORENTINE ARTWORK: Bronze statue of Perseus, with other statues, in Loggia dell'Oreagna, Florence; sunken relief sculpture of a greyhound, housed at the National Museum of the Bargello in Florence; salt cellar with Neptune and Tellus.

famous royal jeweler with the richest clients in the world. By our standards, Cellini ought to launch an IPO, mind his branding, and hire subordinates, lawyers, and a PR firm. Cellini never behaves in any such way. That's precisely why he's a great Renaissance artist: because, by our standards, Cellini is a lunatic surviving on a razor's edge.

Every once in a while, somebody, usually some overfed priest or royal lackey, observes that mere goldsmiths really aren't supposed to be any good at, say, building fortress gates. Cellini despises such dull, insipid warnings. Yesterday's limits don't apply to him. He can make anything.

At one point, while he's jailed in a Roman prison, he tells the warden that he's sure he can make a flying machine. Yeah, that'll be simple. Cellini doesn't fly, but he does escape that prison by secretly hacking the prison door and making a rope from his bedsheets.

In his blazing enthusiasm, this Renaissance man can master a complex new trade in a month. How? Simple: he just doesn't accept human limitations. He buckles down to the impossible task and burns his way through it in a fanatical orgy of work. We'd call that obsessive-compulsive behavior. Cellini has his own terms for it: he claims that he

is "winning renown" and "proving his valor."

Here's Cellini hanging around a palace, with a bunch of overdue royal commissions, as usual. A rival fails to show up for an appointment. Cellini reacts to this insult with his book's most common leitmotif: he grabs a blade. "He refused to come, which made me so angry, that, fuming with fury and swelling like an asp, I took a desperate resolve ... In the fire of my anger, I left the palace, ran to my shop, seized a dagger and rushed to the house of my enemies, who were at home and shop together. I found them at table; and Gherardo, who had been the cause of the quarrel, flung himself upon me. I stabbed him in the breast."

Cellini fails to butcher his rival right at the family's dinner table, mostly because the victim's relatives batter him with pipes and hammers. Cellini takes this minor debacle in stride. A few chapters later, his little brother, who's a Florentine hoodlum as dumb as a box of rocks, gets shot dead in a senseless fight with police. Cellini charges headlong into a crowd of ten Roman cops and fatally stabs the cop who shot his brother.

After picking up the brand-new "French disease" from one of many model-courtesan girlfriends, Cellini breaks out in ghastly fetid blisters. He doesn't bother to ask medical professionals to cure him. Instead, he retreats into the woods, invents a fad anti-VD diet all his own, and "cures" himself. Later on, Cellini also catches the Black Death. The plague only hits him in his left armpit, so it's a minor deal for him. When he's dying of scurvy in a dank, dark papal prison, Cellini pulls out his loose teeth with his own fingers. Dentistry is just another handicraft.

Then Cellini ventures to France to create his greatest works of art. He has to fight off some bandits along the way.

> In his blazing enthusiasm, Cellini can master a complex new trade in a month. How? Simple: he just doesn't accept human limitations. He buckles down to the impossible task and burns his way through it in a fanatical orgy of work.

You'd think that all this disease, urban violence, warfare, and malnutrition would slow down his art production, but it's business matters that really bother Cellini. Cellini never gets paid properly for any stunning masterwork that he creates. Various kings, popes, and dukes promise him the moon, but court functionaries consistently rip him off. This maltreatment doesn't slow his gusting, all-competent creativity. It just annoys him, so that he brutally beats his subordinates. In one minor episode, he kicks a teenage apprentice in the crotch so hard that the kid crashes straight into the King of France.

I'm going on a bit, but these anecdotes are just a few among the dazzling career highlights of this master craftsman. My personal favorite is the extensive episode when Cellini hires a sorcerer to raise a horde of devils straight from hell. The sorcerer actually achieves this feat, and everyone's very impressed. There's also a notable moment when the delighted Pope personally blesses him

for cutting a Spaniard clean in half with a lucky cannon shot.

Certain modern commentators seem inclined to disbelieve Cellini's narrative. They think that Cellini is boasting and bragging. I'm inclined to believe that Cellini is underplaying his experiences. If Cellini had wanted to boast, he would have finished his book and seen it copied and spread around widely. That never happened; his incomplete memoirs weren't published until he'd been dead for 157 years. So Cellini's not boasting. He has the seasoned, even world-weary air of a master craftsman who is trying to get an apprentice up to speed.

He's got a more-than-healthy male ego, but, oddly for a supposed "Renaissance genius," Cellini's not very bright. He's never a fancy-pants intellectual, no complex, tormented Leonardo type. Cellini doesn't much care for philosophy, mathematics, theology, or science. He makes really fine things with his hands; that's what he's great at doing, and that's what life's all about. At heart, Cellini's a regular guy: he likes to ride horses, go hunting, eat, drink, and make merry with women. He also likes to stab loudmouths to death in public, and he may very well be bisexual, but these are technicalities.

What's the big lesson here? It's not that Cellini was a "larger-than-life figure." The lesson is that life is larger than we sometimes think it is. We can easily judge Cellini, because he's dead and we're not. It's a better lesson to wonder how he would judge us.

That answer's obvious: he'd loudly mock us for our sheer cowardice. He'd wonder at our leaden lack of proper dash and spirit. Compared to Renaissance masters — guys with no more than hand tools and maybe a forge and a gear-jack — any artist today has tremendous potential creative capacity, with fantastic access to sophisticated tools, resources, and ideas.

So, why aren't we all Renaissance people, every last one of us? Why are we so numb and sluggish, so feeble and humble? What possible excuse can we offer for our comprehensive lack of awesome glory and grandeur? What's stopping us? Do we imagine we'll live forever? Why don't we just … change everything?

Bruce Sterling (bruce@well.com) is a science fiction writer and part-time design professor.

DRUGSTORE SPECIALS

FROM MAKER TO MAKER

Who doesn't appreciate a really good tip now and then? Especially the kind, as one reader put it, "that changes your life." Whether it's something as practical as investing in a metal punch, or as crafty as using an emery board to smooth out rough edges, we all rely on our friends and neighbors to tip us off to the new and the good.

—Arwen O'Reilly

Vaporware

Maybe it's just an urban legend, but after receiving an email forward about getting rid of coughs by putting Vicks VapoRub on your feet (don't forget socks, so your bed doesn't get all gooey), a friend tried it out and reported back in favor. "It's incredible!" he said in awe. "I've gotta try it on my kids when they get sick."

Pack a Punch

"A lot of people don't realize just how handy a punch is when drilling into metal," points out MAKE columnist Saul Griffith (pun is definitely intended). "Having a dimple to mark the spot makes a huge difference in accuracy. It's a tool everyone should have."

Smart Parts

Eric Wilhelm raves about octopart.com, a search engine for electronic parts. "You enter the part number, or simply its title or use, and Octopart returns the best matched components, their prices and availability from several suppliers, and links to the catalog pages. What's even more exciting is the concept of Octopart searching through a web page, determining which specific parts are mentioned, and generating a personalized shopping cart with the cheapest and most readily available parts."

Tough as Nails

Can't find your deburring tool? Check your medicine cabinet instead. Rachel McConnell, software developer at instructables.com, suggests using an emery board. "You can get lots of different grits, they're stiffer than sandpaper so easier to use, but still flexible, and they aren't as harsh as a metal file," she says. "I've used them on wood, hard plastic, and metal surfaces — usually on metal for burr removal. They do wear out relatively quickly but you can get more anywhere."

Light in Tight Spots

MAKE reader "Monopole" reports: "I was overjoyed to discover at the local supermarket that Gum brand is selling a neat pack of two nice dental picks and a dental mirror with an integrated flashlight in the handle, all for less than 10 bucks! Those are three indispensable tools for any maker, great for looking in difficult-to-reach places. The mirror with the integrated flashlight is worth it alone. When you need a dental mirror, you invariably need a flashlight, as well as a free hand!"

Have a tip for MAKE readers? Send it to tips@makezine.com. Also be sure to check out TNT, our Tools-N-Tips newsletter, at makezine.com/tnt.

ART WORK
Illuminated Circuits

By Douglas Repetto

AN ILLUMINATED MANUSCRIPT IS A TEXT with graphical embellishments: large, ornamented letters at the head of paragraphs; small paintings in the margins or interspersed within the text; borders dense with vines, patterns, and flowers. Sometimes the embellishments serve to clarify or enhance the meaning of the text; other times, the text seems merely an excuse for the intricate, exuberant illustrations. I've been thinking about illuminated manuscripts while looking at the work of a number of artists who are using circuit design as both a functional and an expressive medium. Let's call these designs "illuminated circuits."

The most straightforward way to illuminate a circuit is simply to embellish the functional electrical traces with additional not-so-functional designs. Paul Slocum's *AC 120V Weave* and *DC Power Supply* are both electronically simple circuits made complex by the addition of extra traces in the form of school drawings and book-cover doodles from his youth. The circuits are simply "on" indicator lights, indicating the status of nothing in particular. One motivation for these pieces was Slocum's involvement in the world of "circuit bending." Many circuit benders not only modify the circuits of existing devices, but they also go on to decorate and enhance the original cases. Slocum wanted to focus solely on the idea of decorating machines, bypassing functionality altogether. These works are ornate circuits that do nothing but indicate their own presence.

LoVid (Tali Hinkis and Kyle Lapidus) takes a more involved approach in *Hearing Red*, a warm and buzzy ode to hearing loss. The circuit is a simple video generator that outputs an all-red image. Instead of running the video output into a monitor, it's run through a small speaker so that you hear the static hum of the video sync signal. The rough, wiggly circuit traces were drawn based on the anatomy of the inner ear, focusing on areas that cause hearing impairment when damaged. Some of the traces are functional circuitry, while others are purely ornamental. Twisted and knotted red fabric, also inspired by aural anatomy, is woven through holes in the board. The circuit boards were milled out on a small CNC mill, removing some of the board substrate and allowing light to shine through the traces. To heighten the effect, lights are mounted behind the board, giving it a gentle glow and literally illuminating the circuit's fanciful designs.

The circuits in Rob Seward's *Consciousness Field Resonator* aren't electronically embellished, but the PCBs they're mounted on are hand-finished to give them a warm, aged look. *CFR* monitors the output statistics of random number generators,

> The art and craft of circuit design is an open-ended endeavor, with enormous potential for both pragmatic invention and playful creativity.

lighting up when there's a significant deviation from the expected values. Some experimenters, such as the Global Consciousness Project (*see MAKE, Volume 09, page 62*), claim that there's a connection between highly emotional global events (like the death of a celebrity or a natural disaster) and the behavior of such sensitive electronic circuits. Seward isn't trying to support or debunk that idea; rather, he wants to explore the human urge to make such connections even when they're at odds with

our rational thought processes. The mix of angular circuit traces, softly glowing LEDs, and burnished patina make *CFR* look like it's out of the steampunk version of a Victorian natural history museum.

In addition to embellishing circuits, some artists ask questions about the basic physical form of the circuits themselves. Peter Blasser's Rollz-5 drum machine schematics (available for free at ciat-lonbarde.net/rollz5) are meant to be printed and built on paper, with the component outlines printed on one side and lead connections on the other. You poke holes in the paper at the indicated spots, insert the components, and solder away! The transistor-based pulse generation circuits that provide the drum machine's rhythms come in three, four, five, and six-node versions, and the number of nodes determines the visual shape of the circuit itself — three nodes make a triangle, four a square, and so on. The circuits and instruments Blasser designs are always functional, but their expressiveness comes as much from their playful visual appeal as from the sounds they make.

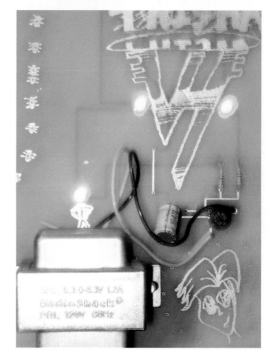

Peter Vogel's interactive light and sound works and Michael Ang's *Blue Flower* sculpture use electronic components as both circuit elements and physical structure. The etched-copper petals in *Blue Flower* function as power and ground rails, and the LEDs that run off those rails do double duty, both lighting the flower and holding the parts of the petals together. Ang's elegant design is purposefully reminiscent of biological "design," where each part of a flower serves a purpose. Vogel's wiry constructions are much more industrial-looking; component leads are joined together to form columns, spirals, and physical supports for speakers, sensors, and LEDs. His light-sensitive piece *Sounds* is like an electronic Mondrian painting, colorful components arranged in interlocking geometric forms.

The art and craft of bookmaking have proceeded at a tremendous pace since the medieval heyday of illuminated manuscripts, as materials, techniques, and even motivations continually evolve. I imagine a similar trajectory for the art and craft of circuit design, an open-ended endeavor with enormous potential for both pragmatic invention and playful creativity.

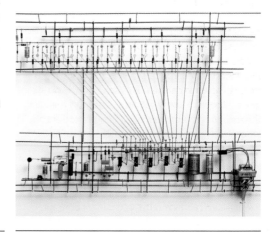

Douglas Irving Repetto is an artist and teacher involved in a number of art/community groups including Dorkbot, ArtBots, Organizm, and Music-dsp.

From top left: LoVid's *Hearing Red*, Michael Ang's *Blue Flower*, Paul Slocum's *DC Power Supply*, Peter Vogel's *Moving Lights*.

Downhill Makers

Garage ski builders are outdoing the pros. By Jason Verlinde

SOME MODES OF TRANSPORTATION WERE destined for garage builders. Think of the thousands of wooden canoes, dune buggies, and steel-framed road bikes that hobbyists have created over the years. Modern, high-performance downhill skis, on the other hand, seem to be an entirely different beast, the kind of state-of-the-art product that only a big factory production process could churn out. Who else has the ability to fuse together all those exotic materials into a sturdy package that will safely get you down icy slopes, powder runs, and even the occasional cliff huck?

Well, it turns out that virtually anyone with basic woodworking skills can make a pair of skis. And the three friends behind the skibuilders.com site — Kelvin Wu and cousins Kam K. Leang ("Big Kam")

and Kam S. Leang ("Little Kam") — are hoping to inspire other snow fanatics to build their own rides. "The vision of SkiBuilders.com is to demystify the design and engineering of skis," Big Kam says. "It's a web-based information clearinghouse for people who love to ski and build things."

THE BUILD
In Wu's Seattle garage, he and Big Kam walked me through the ski building steps. To start, you need some common power tools. A sander, a router, and a table saw are all essential. You'll also need to fabricate two important items: a core profiler and a ski press. The core profiler allows you to trim and taper the laminated wood guts of the ski, something that would be impossible to do consistently by hand.

Photography by Jason Verlinde

And the ski press — either pneumatic or vacuum — compresses and heats the skis into their finished form. (Plans for all of the above, and links for part suppliers, are available at skibuilders.com.)

First, you use your profiler and a router as a guide to shape your wooden core. Poplar, maple, and birch are all popular woods to make skis from, but you can be creative in the wood selection and in how you layer each type — the finished laminate looks like a thin kitchen cutting board in the shape of a flattened ski.

Basically, you make a snow-ski lasagna, with fiberglass layers acting as your noodles.

Next, you cut your ski base material to the desired ski shape, and bend edge material around the contours of the ski; at this point, a thin strip of rubber is often added to the top of the edges for vibration dampening. Now we get to the fun, and messy, part: epoxying layers of fiberglass sheets to the wood core, base, and edges. Basically, you make a snow-ski lasagna, with fiberglass layers acting as your noodles. You can also go wild with your graphics (the sample pair shown on the next page features a cut sheet of fabric and some printed paper logos) and place them under a clear topsheet (often made of clear P-Tex, the same plastic used for the bases).

The ski press that Wu built for his garage is pneumatic: two ski-length sections of 5-inch fire hose are clamped off and filled with compressed air to force pressure onto the skis (a method taken from skateboard deck manufacturing).

Here, pushed against a heated ski mold for an hour or two, the pair will cure into their desired shape and camber. Some garage builders have simpler setups: single ski presses made out of wood and car bottle jacks. Basically, you just want to safely get the skis to stick together long enough for the epoxy and fiberglass to become solid.

When the skis come out of the press, they are often attached to each other by the now-dried fiberglass and epoxy hanging off the edges. Using the skis' edges as a guide, you use a jigsaw to cut off the excess material, and then use a router to shape

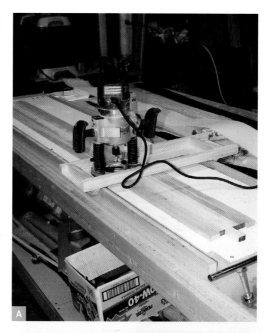

Fig. A: A router shapes the wooden core.
Fig. B: A thin strip of rubber on top dampens vibration.
Fig. C: Wu's pneumatic ski press.

the sidewalls and do some final touch-up work with a file. From there on, it's the same as with store-bought skis: bindings have to be mounted and a base grind and tuneup will be needed to get them performing at their peak. The entire process (minus the binding mounting and tuneup) can be done in a long day.

As complicated as it all sounds, by your second pair of skis, you may already be saving money compared to buying high-performance skis off a rack.

"Costs to get this setup can vary dramatically, but depending on your scavenging skills and ingenuity, a setup can be as little as a few hundred dollars," Big Kam says. "We highly recommend that first-time builders try to scrounge up as much material (free or discounted) as possible to lessen costs.For the tooling, which includes pneumatic press, core profilers, sanders, routers, et cetera, I would estimate the total cost to be near $600. The material to build one pair of skis can be as low as $50."

To date, the guys behind skibuilders.com have added about one pair of skis a month to their site's ski gallery. They're now up to almost 30 pairs. There are reverse cut and reverse camber creations, traditional-shaped skis, twin tips, and swallowtails.

Most are either based on existing, off-the-shelf skis, or are exotic, flight-of-fancy creations. Making them quickly and testing them is, at least for now, more important than long-term durability. In fact, each member of the site has a good delamination story.

"Getting things to bond well is a tricky process," Little Kam says. "Last season I was skiing on a pair of skis on which I tried a new method of using fabric paint for the graphic. After a few icy landings in the park the topsheet completely ripped off ... the paint separated. I decided to ski with the topsheet flapping around for the rest of the day, much to the confusion of Southern Californian snowboarders."

Wu admits that the group still has a lot to learn about the building process. "There are a lot of material treatment and bonding issues that we need to figure out, and how best to do it in a garage shop," he says. "I think most of our delamination could be eliminated with proper treatment and cleaning. Also, there isn't a lot of empirical data on how the flex, sidecut, length, and other aspects of ski design work together and how they affect the ski's performance. We also still need to figure out how best to control the camber of the skis in a heated press. It seems that the heat and the

Fig. D: When they come off the press, the skis are attached to each other by the fiberglass sheet.
Fig. E: Using a router to shape the sidewalls.

differences in thermal expansion of the materials make it really hard to predict how the ski will turn out once everything cools down."

THE TEST

After spending a day on a pair of garage-built skis, I was hooked. I can honestly confess that they performed as well as the all-mountain, high-end K2s I've used for the last two seasons. The garage skis garnered a lot of strange looks from the lift attendants but they were an absolute blast on the runs. Granted, I was worried the whole time that they might delaminate in the middle of a steep run, but they never did. And, even if they do break, I now know how to make more!

Seattle-based Jason Verlinde usually writes about guitar, not ski, builders for his magazine the *Fretboard Journal* (fretboardjournal.com).

Make:
technology on your time™

BUILD COOL THINGS!

4 Quarterly Volumes for only $34.95—SAVE almost 42%

Name (please print)

Address   Apt.#

City   State   Zip

Email Address

MAKE will only use your email address to contact you regarding MAKE or other O'Reilly Media products and services that may be of interest. You may opt-out at any time.

4 Volume Rate: $34.95. Savings based on $14.99 cover price.
Canadian Rate: $39.95 USD (includes GST). All other countries: $49.95 USD

For faster service, please order at **makezine.com/subscribe** and use promotion code **B7A6A**

makezine.com

Lucid Dreaming Mask

Timed LEDs prompt your brain to direct your dreams.

By Nathan True

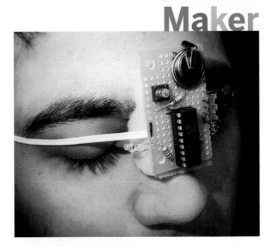

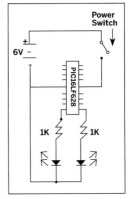

YOU'RE SITTING IN YOUR CAR, DRIVING to work. At a stoplight, the car across the way starts flashing its lights at you. Squinting, you think: *What's this guy's problem?* Lazily, you recall something about bright lights ... and then you remember. *Flashing lights mean I'm dreaming!* You take a moment to confirm it (yes, your glove compartment is filled with goldfish, as expected), then step calmly out of your car and decide to fly through the air.

This is the "lucid dreaming" state, which lets you interact consciously with your dream worlds and break the rules of reality. Lucid dreaming is fun, and enthusiasts have developed many ways of trying to induce the phenomenon, from simply repeating statements of intent ("I *will* realize I am dreaming tonight") to using hypnosis and brain wave analysis.

Technological solutions have also emerged, the most noted of which is the Lucidity Institute's NovaDreamer, a product that uses infrared light to sense the rapid eye movements (REM) that characterize the brain state associated with dreaming. Once it senses your REM, the NovaDreamer cues sounds and flashes lights into your eyes. This injects the experience of flashing lights into your dream, and with luck (and training), you will recognize the lights, realize that you're dreaming, and be able to start playing your own first-person dream-world game.

Unfortunately, the NovaDreamer and similar products are expensive, costing $200 and up. Many DIY efforts (most notably the Kvasar, online at brindefalk.solarbotics.net/kvasar/kvasar.html) have attempted to copy the commercial masks, but these weren't reliable enough for me — they require tight-fitting masks to sense REM properly, and often need careful calibration to detect anything at all.

When I thought about it, I realized the masks were

more complicated than they needed to be. If I'm not dreaming, what would flashing lights do? Nothing at all — they would just flash and I would keep on sleeping. Thus came my epiphany — I could just set the lights on a timer and have them flash at intervals once I've fallen asleep. With this design, the hardware is drastically simple: LEDs pointed at the eyes connect directly to a PIC chip. The only other components are a battery, a power switch, and a mode button.

The program inside the PIC is dead simple, too — it waits 4 hours, then flashes the LEDs in a predetermined pattern (5 flashes, pause, 5 flashes) every 5 minutes. The LED brightness is carefully controlled so as not to wake you up, but still bright enough to make it through your eyelids and into your dream.

Using the mask is as simple as turning it on and strapping it to your face. It's surprisingly comfortable, and falling asleep with it on is no problem at all. Then just keep an eye out for flashing lights!

I've put together Lucid Dreaming Mask kits, which I sell for $30 each (or $10 for the programmed PIC alone) at lucid.natetrue.com.

Nathan True is a casual inventor, tinkering in both hardware and software. He runs the website cre.ations.net.

Photography by Nathan True

JUNKYARD DOG: Molly, shown here, knows the importance of keeping amassed junk organized, especially when battling unfruitful failures.

Fail Early! Fail Often!

A mental toolkit to sharpen your skills. By Tom Jennings

NO ONE TALKS OF FAILURE AS ANYTHING but shameful; this is wrongheaded and foolish. Mistakes are synonymous with learning. Failing is unavoidable. Making is a process, not an end. It is true that deep experience helps avoid problems, but mainly it gives you mental tools with which to solve inevitable problems when they come up.

It all begins with a mental toolbox, filled with useful items you can't buy, but can only obtain through the act of failing again and again. Here are mine.

TOOL: The Dorkifier

You may think that in order to look cool to your peers, you must never look foolish. Abandon this. Cool is precisely the opposite of pursuing a project that taxes your brain and body, and might not even work! Fear of looking like a dork ("You're building a what?") stops a lot of people. Give up any thought of looking good, and instead make good.
CONCLUSION: Embrace dorkiness!

TOOL: The Troubleshooter

Since roadblocks and failures are a given, how to proceed? This is the key to all project making: troubleshooting, problem solving, debugging.

Don't freak out. Become methodical, or contemplative, or go get a beer. You need your brain in its best working order.

Troubleshooting really means: what do I have to learn to resolve this problem? This is true if the problem is reviving a dead motor or using glow-in-the-dark paint; you read, Google, practice, experiment.

Bob Pease, god-king-emperor of analog electronics

Photography by Tom Jennings

and author of *Troubleshooting Analog Circuits*, asks: "Did it ever work right? What symptoms tell you it's not working right? When did it stop working, or start working badly? What other symptoms showed up just before, after, or during the failure?"

CONCLUSION: **Problem solving is the universal solvent.**

TOOL: **The Perspective Rotator**

Sometimes "failures" are really successes. Really. I use a very temperamental chemical process to make deep-etched metal panels. It's very flaky, and perfection is nearly impossible. But I soon realized that the flaws and corrosion had beauty in themselves, and actually improved some of my artwork.

CONCLUSION: **Making is a process; adapt!**

Water ruined this printed label ... which looks appropriate on this very old teletype.

TOOL: **The Opportunity Multiplier**

Sometimes failures really are failures. Do not give up, for there is always another project. Always! For every successful project, there are ten ruined hulks to go in the junk box.

CONCLUSION: **There is an infinity of projects.**

TOOL: **The Autopsy Kit**

You may battle some device or technique, swearing, and sleepless, your spousal unit calling you to bed, you too bitter to do anything but stew in your juices, all because you realize you cannot affix Part A to Part B until you have first affixed Part B to Part A ... or more subtly, It Will Never Look Right.

Despair, certainly, but do not give up. When you have cooled, do a post-mortem. Imagine an autopsy on a TV cop show. A good post-mortem asks the right questions: not so much "What's wrong with this thing?" as "Why can't I fix it?"

Mistakes and failures don't make you a bad person. Resist the Western good/bad, fail/succeed binary. Fear of failure can devolve into macho posturing (one antidote is dorkiness).

Last and not least, write down problems and failures! Bob Pease quotes Milligan's Law: "If you notice anything funny, record the amount of funny." You'll be amazed how useful this will be later.

CONCLUSION: **Do post-mortems; take notes.**

Big oops: Hole for A/C duct in wrong place! Ouch!

A year's worth of failed attempts to make a clear LED taillight lens!

Vacuum tube analog computer nuclear reactor simulator, failed: too ambitious! Maybe someday ...

Maker

TOOL: The Persistence Enhancer

Once you start failing on a regular basis, you'll learn the value of persistence. If your project is a challenge, you will have problems. It's unreasonable to expect quick results, ever. If you have to stay up all night, so be it. It might take you a day, week, month, or year to gain the skill(s) to complete something.
CONCLUSION: **Persist!**

AVOIDING UNFRUITFUL FAILURES

While failure isn't necessarily a bad thing, there's no reason to invite it into your life with open arms. There are two kinds of failure: meaningful ones, which add more tools to your mental toolbox (see previous), and irritating ones, which threaten to damp your enthusiasm and interest. To prevent the latter flavor of failure, here are a few things you can do:

1. Get out of a rut.

Makers are generalists, with a broad range of skills. As a generalist you will find that your skills and brainpower in other areas improve when you learn any new skill.

Doing one task for too long leads to mental fatigue. Also, your brain actually digests experiences after you stop doing them. Those two facts synergize. Wrote code for two days straight? Do some gardening! A long day of woodcarving? Do some writing! Achieve balance by practicing extremes. (My most common pairing is software and electronics vs. automobiles and gardening.) Hanging out with animals is an excellent grounding experience.
CONCLUSION: **"Specialization is for insects."**
—*Robert Heinlein*

2. Collect a critical mass of junk.

The truly wise have deep and rich junk boxes: a fistful of choice tidbits in a shoebox under your bed, or an airplane hangar with kilotons on ceiling-high racks.

If you can't find it, you might as well not have it.

There is a cultural element to junk collecting. Some relish a yard full of old car parts (that's me). At the university where I work, we took a field trip to a local industrial surplus store, and one young student from a non-Euro culture was visibly disturbed to see professor, staff, and students gleefully climbing on filthy, wobbly piles of junk (treasure).

Junk collecting is, to me, a discipline; it requires physical and mental effort. Will this (random) thing help me in my project, or am I mesmerized by this shiny bauble's beauty? Does it fit? Can it be made to fit? Can I adjust my project/path/goal to accommodate it? Does it suit my aesthetics? My politics? Can I afford it? Can I afford to pass it up?

Useless beauty! Who could resist?
Failed brackets and parts: fodder for future projects.

I work best when my junk is distilled to exquisite perfection. I have more space than many people, but hardly infinite; when I bring something home, something else must go. I have been doing this for more than 20 years. The result is that nearly none of my junk is, well, junk, at least not to me. Remember what I said about being a dork?
CONCLUSION: **Good junk is good.**

3. Seek out the old-school engineers.

Never ask engineers how to build things, unless they are old people. Today's engineers sit in cubicles and type on keyboards. Old people actually built things, like designed an amplifier, chose the parts, made the cabinet, wired it, and then sat back and listened to it. Today, young engineers will snicker at your harebrained projects.

But engineers are not stupid; it's just that engineering is not craft. Fordism removed skilled craft from capitalist projects a century ago. Engineers are taught to engineer, not to build.

When you design a circuit for your project, you will likely make one, or a few, of them. Commercial products produced in the millions are engineered for lowest possible cost, and fewest returns to stores by customers. Ten more cents' worth of copper matters when you make 50 million iPods. Ten feet of extra copper wire matters not at all when you are making a Tesla coil to ruin your neighbor's TV reception.

The craft of electronics requires a different set of techniques than engineering does, though clearly the underlying physics of electronics is the same. The same is true in all skill areas.
CONCLUSION: **Engineering is not craft.**

4. Take one step at a time.
You should challenge one or two skill areas in a project. If you know how to do electronics and PIC programming, say, then go make that mechanical robot arm. But if you've never done electronics or programming before, that might be a foolish project; build a simple electronics kit first. Remember, making is a process!

When (not if) you are a neophyte in some skill, it's hard to know how to begin. I have decades of experience in electronics, software, and metal working, but no idea how to use plaster or model railroad materials. My experience in other areas gave me the confidence to ask stupid questions at the model store, and buy stuff to experiment with. Now I have the leftovers on the shelf for future projects, and the beginnings of a new skill!
CONCLUSION: **Be reasonable!**

5. Plan ahead — visually.
I'm shocked when I see people start projects without the slightest bit of written or drawn visualization. Art-trained people are better at this, keeping notebooks filled with scribbles, notes, and drawings. These are critical thinking tools. They are also communication tools. Others cannot read your mind; if you cannot document your ideas, no one will understand you. And chances are, a few years from now neither will you. Computers are often very poor tools for this; you were warned.
CONCLUSION: **Write! Draw! Document!**

6. Find inspiration in everything you do.
Sometimes the urge to do something hits me, and must be satisfied, and no current project is conve-

nient. The solution? Putter! Organize your car parts. Unshelf/reshelf all the paper-tape gear. Peruse obsolete catalogs. Put another coat of shellac on that radio. All of these things are inspiring and worthwhile. You might even rediscover some forgotten artifact and embark on a new project!
CONCLUSION: **Puttering is good.**

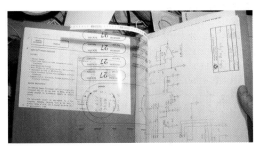
Write and draw in pencil, and note changes!

Keep project notes together.

Keep notebooks together.

I hope I've made it clearer that failures and problems are not character flaws or reason to quit, but are how we all learn new skills. Making is about the process more than the artifacts; it's about remembering that a mental toolbox is the most valuable, and the most portable!

Tom Jennings (tomj@wps.com) is a technical artist and chronicler of dead technologies living in Los Angeles.

"Wealth Without Money"

A machine that can make almost anything, including copies of itself. By Matt Sparkes

QUIETLY AND DILIGENTLY, IN LOCATIONS all over the world, an organization is working to bring the means of production to the masses. Their motto, "Wealth without money," implies a political motivation, though they're not socialists, Marxists, or communists — they're hardware hackers.

Adrian Bowyer, at Bath University in the U.K., leads the RepRap project (reprap.org) to develop a design for a very special rapid-prototyping machine. The device treads a fine line between being capable enough to produce complex goods, and simple enough in design that it can produce its own parts. The result is a self-replicating machine, with the potential to spread across the planet at an exponential rate.

The RepRap (Replicating Rapid-prototyper) is a fused deposition modeling machine that extrudes molten plastic or metal from a heated nozzle. It works in an additive fashion, building up an object by depositing layers one by one. The nozzle is controlled by a computer, which moves it in three dimensions to create the necessary shape.

The machine's replicating ability alone is enough to provoke interest in the project, but the really interesting part is what could be made with it — and Bowyer has some potentially revolutionary ideas. "Making complicated things is not a large step," Bowyer says, "certainly not as large as the step from making nothing to making simple things."

For example, RepRaps could someday build an open source mobile phone network by creating

Photography courtesy of RepRap

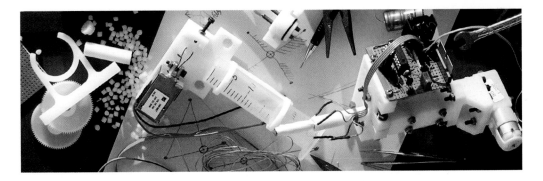

An early experimental polymer extruder, with assorted components and materials.

phones that also act as base stations, routing calls from peer to peer. This free communication would have a phenomenal impact on developing countries. Another possibility is DIY medicine; as drug patents don't apply to personal production, expensive patented medicines could be saving lives worldwide if the machines could be made to produce them.

You would be forgiven for thinking that a project like this would be incredibly high-tech and well-funded, but there is a strong amateur element, composed of enthusiastic makers. The RepRap plans and software are freely available under a GNU license, and participation in the project is open to anyone with the skills and motivation. The benefit of a project that aims to create a simple machine is that the barriers to entry are very low; you could build your own machine and start to refine and improve components, or help to develop and improve the software. There is even demand for help with the project documentation.

Zach Smith is a web developer by trade who took a few electronic engineering courses in college, and is building his own machine. "All I know about the circuits, I found out by asking questions and looking online," he says.

Forrest Higgs, whose background is in aerospace engineering and architecture, has one too, and he's trying to refine the concept by deviating from the main plans. He proclaims to be on the "bleeding edge," and any improvements he makes will be fed back into the project.

Some parts to create the machines are currently bought, rather than produced by a parent machine. "We buy nuts and bolts, as it's cheaper and easier. It would be nice to have a pure machine, but I would imagine that it would be some time before these readily available parts are produced by the machine itself, even when they are possible to be made," says Bowyer. Some of the home-builds even use Lego, Meccano, or wood to build the chassis.

According to Bowyer, within two years the RepRap machines could be producing all their constituent parts, even electrical circuits. "There are two alloys, Wood's metal and Field's metal, which have lower melting points than that of the plastic used in the machine's construction. We would have a small chamber that heats this alloy, and we can then extrude this from the same nozzle we use to extrude the plastic. The big advantage is that you can then put a layer of plastic on, and another layer of alloy — you have 3D circuits. You can then hide these circuits in the actual structure of the machine."

The controlling computer will probably be pre-produced for the foreseeable future. The software is currently written in Java because of its platform

> "Making complicated things is not a large step, certainly not as large as the step from making nothing to making simple things."

independence, but the team is closely watching MIT's $100 computer project (laptop.org); it would provide a perfect controller, is affordable, and could use the same 12-volt power source as the RepRap machine.

RepRap has the potential to do something really revolutionary, to bring about a manufacturing singularity with a very real impact on people's lives. The idea that these machines can reproduce and evolve is fascinating. The hackers who make up the RepRap project will constantly improve the design, making it simpler to make and more capable. What better outlet is there for your maker tendencies?

Matt Sparkes lives at mattsparkes.org, and also in London.

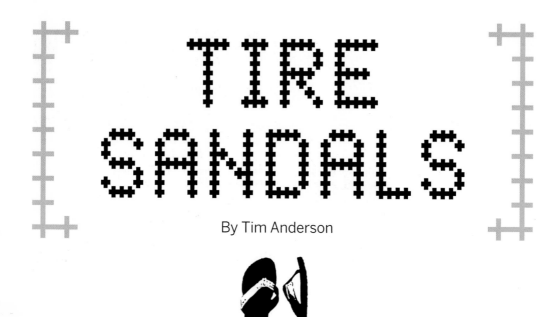

TIRE SANDALS

By Tim Anderson

Solve multiple global problems at once when you make your own sandals from an old tire.

An accumulation of old tires is an increasing problem in the United States. Tires can't be easily recycled, and mosquitoes breed in the water that collects in them. But you can solve multiple problems at once when you make your own sandals from an old tire.

Rubber tires make great shoe soles. The line "huarache sandals, too" in the Beach Boys song "Surfin' U.S.A." refers to the Mexican tire-soled sandals once worn by well-dressed surfers. The Viet Cong and other armies have been shod almost entirely with such footwear.

Once found in every country warm enough to wear sandals in, they're increasingly hard to find. The last time my dad tried to buy huaraches in Mexico they had injection-molded soles instead of the traditional tires. He was told that Mexicans associate the tires with poverty. So it looks like we'll have to make our own. I prefer flip-flops to sandals, so that's what I usually make.

TIRE FLIP-FLOPS

I've been wearing a pair of these for a couple of years now. I've worn them in Costa Rica, Nicaragua, China, and all over the U.S., and they're still going strong. They seem stiff compared to the usual flip-flop, but that becomes a virtue when walking on sharp rocks that would destroy ordinary flip-flop soles.

Now that I'm used to them, it's far more comfortable than going barefoot, even indoors. The curve from the tire tread is nice — it helps keep the shoe on your foot and it gives your stride a bit of a "rolling" feel.

NOTE: Don't use a steel-belted tire. It can be hard to find a non-steel-belted tire in the U.S., but spare tires often have fabric belts rather than steel. Most of the world can't afford steel-belted tires, so it's not a problem outside our borders.

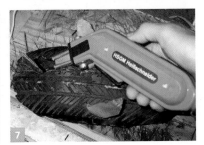

1. Cut the tire rubber.

Here's the easy way to cut the rubber: if the area being cut is under tension, it's easy to cut with a slashing or sawing motion. Use a rough stone or sandpaper on your knife so it has a nice ragged edge. Resharpen as needed.

After you wear your sandal for a while, you'll see where your foot rides, and can then cut away unnecessary rubber from the margins (shown above).

2. Make the instep strap.

This is what the first stage of the foot strap assembly looks like. A seatbelt folded over to make it half-width works well, but try other materials also. This one is made from a chunk of nylon dog leash. Melt the ends after cutting so they don't fray.

3. Sew the instep straps.

Use a zigzag stitch back and forth over the joint between the two chunks of strap. Go back and forth until the two are firmly joined. Most consumer sewing machines will handle this job without too much trouble.

4. Sew the toe strap to the instep strap.

If your home sewing machine does have trouble, you can sew it by hand. This is what the strap assembly ends up looking like.

5. Cut slots for the straps.

Use a heavy hammer and a chisel or other sharp object of the appropriate size. Put a board underneath to spare your floor. The proper angle is shown above.

It's tempting to cut slots parallel to the edge of the sole. Don't. They won't grip the straps well. Also the straps won't lay flat against your foot. The edge of the strap will chafe you and give you a sore.

6. Poke the straps through the slots.

Shove the straps through the holes with a screwdriver. Or pull them through with needlenose pliers. The harder it is to do this, the better the strap will stay where it belongs.

7. Melt the ends.

Here I'm using an electric "heißschneider" (German: "hot cutter") to cut the end of the toe strap and melt a big blob on the end. If the strap exits in a groove in the tread, you can work the blob down into the tread. Any metal tool or heavy scrap metal heated in a flame works just as well as this official-looking tool.

Done!

Enjoy your new footwear. Festoon them with tassels and flashing LEDs. Always be on the lookout for suitable tires and strap material. Teach all your friends to make them!

➕ Go to makezine.com/10/heirloom to learn how to make traditional Vietnamese tire sandals and see cool photos of Samburu tribesmen wearing "kilikili" tire sandals in Kenya.

Tim Anderson, founder of Z Corp., has a home at mit.edu/robot.

Photography by Tim Anderson

Burn to Learn

The Crucible industrial arts school's community of practice.

By David Pescovitz

Photography by Steve Double

THE STORY UNFOLDING ON THE STAGE COULD only be described as Romeo and Juliet go to hell. In January, California industrial arts school The Crucible reinvented Shakespeare's classic as a spectacular fire ballet danced against a backdrop of burning scenery.

Inside a huge Oakland warehouse studio, Romeo and Tybalt dueled with flaming swords on a multi-level stage set. A dervish spun across the stage with his skirt hem spewing fire. The performance, driven by classically trained ballet dancers, breakdancers, aerialists, and even martial artists, was white-hot.

During the Capulets' ball, all eyes were on the daring acrobats twirling overhead from a massive, flaming chandelier. Suddenly, a spray of water burst from the ceiling. Illuminated by the ghostly blue stage lights, the mist created a gorgeous magical glow above the ball below. Then, a scream from one of the acrobats revealed that this watery climax was not planned: "Get me down!" the dancer yelled.

It turned out that the propane-fueled chandelier had triggered the industrial-strength sprinkler system, unleashing a flood of water across the stage and the first three rows of the audience. After the dancers were safely lowered to the drenched stage, a figure in a leather cowboy hat and long black coat walked onto the stage. This was Michael Sturtz, executive director of The Crucible, who had charmingly but awkwardly delivered the play's prologue. The audience burst into applause and with a flourish of his coat Sturtz bowed, seemingly to both the unexpected chaos and also the acclaim of his capacity crowd.

There were no boos. No angry patrons demanded their money back. Dozens of volunteers scrambled into action, mopping up the stage with each and every available rag. Meanwhile, the Oakland Fire Department kindly reset the sprinkler system to a standing ovation. Finally, after more than an hour, the show went on.

Romeo & Juliet: A Fire Ballet was The Crucible's 8th anniversary fundraiser. And in some ways, the opening night performance, including the watery interlude, embodied the essence of The Crucible.

"The word *crucible* has three definitions and all of them fit what we do here," Sturtz says. "A crucible is a vessel used for melting substances at high temperature; the second definition is 'a test of belief or patience'; and the third is a place where concentrated forces come together to cause change or development."

Founded in 1999, The Crucible is a nonprofit oasis of industrial arts education located in the tough waterfront neighborhood of West Oakland. The vast 56,000-square-foot studio hosts 500 classes each year on arts and crafts as eclectic as welding, jewelry, neon, blacksmithing, woodworking, kinetics, and, yes, fire performance.

Last year, more than 5,000 students got their hands dirty at The Crucible, learning how to express their creativity by shaping, pounding, cutting, and molding steel, ceramics, fabric, enamel, and dozens of other media. Another 3,000 schoolchildren, the majority from the surrounding community, participated in The Crucible's mostly free youth programs.

"The Crucible is all about accessibility," Sturtz says. "Anyone can take a class here and almost anyone can teach here too. You don't need a master's degree. You don't need anything other than a passion for teaching, experience doing something, and the desire to share what you know."

THE FIRE BRIGADE:
Michael Sturtz, founder of
The Crucible industrial arts
school, stands with one
of the group's fire trucks.
It is actually a fully mobile
classroom, often seen at
Bay Area festivals.

Indeed, The Crucible is a community of practice, a hub for social learning where individuals gather to share ideas, learn from one another, and create things. These communities of practice are increasingly popular ways for makers to meet one another, find mentors, and tackle projects that verge on the impossible for any one person (see sidebar on page 46).

Born and raised in the Bay Area, Sturtz is the son of an orthopedic surgeon and was expected to follow in his father's footsteps. "I was the child who didn't pass out watching dad work in the emergency room," Sturtz says. Yet after his parents divorced, he also gained inspiration from his mother's boyfriend, the owner of an auto body shop.

"Those seem like different worlds, but both are about repairing bodies and involve mechanics and, actually, similar tools," he says.

Ultimately, Sturtz pursued his creative passions, studying at a handful of art colleges and landing in a West Oakland warehouse loft after graduation. The neighborhood was something of a hotbed for sculptors at the time, although few knew each other. This disconnect reminded Sturtz of what he hated about art school. "We used to say there were too many artists and not enough people." Competition, he explains, can quash community. Now in the real world, he decided to take matters into his own hands.

"Driving by, I'd see a stack of metal or a sculpture outside a warehouse," he says. "So I'd knock on doors and talk to people. Eventually, my studio became a meeting place. I'd have bronze-casting weekend barbeques and invite everyone I knew in the area."

At the time, Sturtz was also teaching at various schools around the Bay Area. Eventually, his classes became "too big and too industrial" for the facilities, so he began teaching out of his own studio. The seed for The Crucible was planted.

Several years later, armed with a $1,750 grant from Levi Strauss & Co., Sturtz and his friends set up shop in an industrial space in Berkeley. After just three years, zoning battles forced the group to find a new home. Welcomed with open arms by Oakland Mayor Jerry Brown, the facility ultimately relocated to its current West Oakland location, a former pipe warehouse. A large portion of the building's sale price was donated by the sellers, a family who was moving their cardboard tube business to bigger digs.

Supported by tuition, donations, and fundraising events like the fire ballet and an annual Fire Arts Festival, The Crucible's unbridled independence is unparalleled by traditional art schools. Where else could one take a class called "Cold Flesh and Hot Metal," with live nude models posing in the foundry?

"There's a commitment to interdisciplinary interaction and a fluidity that enables new programming

> "The industrial arts are dwindling, but a lot of people are interested in learning these things. It's about accessibility to the tools, the process, and the teachers."

to be envisioned and created very quickly," says Mary White, codirector of the glass arts program.

For example, The Crucible is planning a hot glass shop likely powered by solar panels on the roof and a biodiesel generator. As a baby step in that direction, the Crucible staff recently rigged their own ingenious plumbing system for cold working glass. Classes can commence in just six months, demonstrating "an agility that's just incredible for an art school," White says.

For most of the course offerings, The Crucible is the only game in town. (In fact, the school attracts students from all over the region and far beyond.) When the school opened, they expected to provide job training for the area's metal shops, some more than a century old. Now, Sturtz's office is decorated with antique wood office furniture and vintage signs donated by the companies as they shut their doors, hammered by cheap overseas competition.

"The industrial arts are dwindling, especially with the funding for school art and shop programs disappearing," Sturtz says. "But a lot of people are interested in learning these things. It's about acces-

VIEW FROM THE INSIDE (clockwise from top left): Crucible founder Michael Sturtz, on the biodiesel motorbike that recently hit 121 mph at a land speed competition; a bronze pour in one of The Crucible's classes; busy in the blacksmithing space; a welding project; the crucible itself; and students in the woodworking space.

sibility to the tools, the process, and the teachers."

For example, the Bike Fixathon draws dozens of young people who work with volunteer bike mechanics to learn how to fix their own rides. For kids without wheels, there's the Earn-A-Bike program with each participant repairing two bikes, one to keep and one that's sold to support the program.

Coming soon is an art bike program where kids will design, cut, and weld their own choppers, tandems, and other unusual human-powered vehicles from scavenged parts. The Crucible's adult students range from artists to retirees to lots of folks "who always wanted to weld but have no real reason why," Sturtz says.

"We also get a lot of engineers and architects who design in metal or wood but have never really touched the materials."

While his office garb remains work shirts and steel-toed boots, Sturtz admits that he's now doing more administration than he ever thought he'd be capable of back in the days of his bronze-casting barbecues. Right now, he's procuring donations of money and equipment for the new hot glass shop. He's also advising organizations as far away as New Zealand and Ireland on how to create their own Crucible-like communities of practice.

Yet, he still embraces every opportunity to roll up his sleeves. The ballet's flaming chandelier was his creation, and he oversaw the conversion of a 1960 fire truck into the Educational Response Vehicle (ERV) complete with a full metal shop. At the first Maker Faire, the ERV blew away visitors with 30-foot flames belched from its onboard flamethrower. Most recently, he led a team that built a veggie-oil-fueled racing motorcycle. Funding for the bike came mostly from Sturtz's own home equity loan.

"It just had to be built and nobody was doing it," Sturtz says. "Many people are tired of just being consumers. There's a renaissance of people who want to build things themselves. Places like The Crucible demystify how to do it."

More information: thecrucible.org

View a photo slide show of our visit to The Crucible: makezine.com/10/proto

---

MAKE Editor-at-Large David Pescovitz is co-editor of boingboing.net and a research affiliate of the Institute for the Future.

Hot Industrial Arts Around the Country

Wish The Crucible were closer to you? Well, something similar just might be. There are a number of fine institutions around the country teaching everything from blacksmithing and welding to glassblowing and fire dancing.

The Steel Yard in Providence, R.I., is probably the closest analog to The Crucible. With class titles like Introduction to Bladesmithing, Welding with Mom, and Basic Bike Maintenance, it's enough to make anyone's heart beat a little bit faster. They're all about "spreading the industrial art love," so go get some. thesteelyard.org

While the Brookfield Craft Center in Connecticut has a broad and ancient mission ("to stimulate interest in handmade objects of good design"), you can still learn how to turn wood, make books, weld, and blow and fuse glass. There should be enough to wet anyone's whistle. brookfieldcraftcenter.org

In 2004, Holly Fisher was awarded a $100,000 grant from Michigan's governor as part of the "Cool Cities Initiative." Smartshop was the result, with a wide variety of metalworking classes, and classes on building everything from a garden trellis to your own workshop. smartshopkalamazoo.com

While aspiring blacksmiths may have to look elsewhere, glassblowing fanatics will find plenty to keep them busy at Art by Fire in Seattle. And if you really want to play with metal, try electroforming class. artbyfire.com

With a staggering 830 weeklong or weekend classes, the John Campbell Folk School in Brasstown, N.C., teaches everything from lace-making to blacksmithing, basketry to surface design, and even broom-making, chair seats, and storytelling. folkschool.org

Know of one that's not on our list? Send us an email at toolbox@makezine.com.

—*Arwen O'Reilly*

CRUCIBLE SPECTACULARS:
TOP: The Crucible's Fire Arts Festival takes place every July.
BOTTOM: Things heat up at the Capulets' ball in *Romeo & Juliet: A Fire Ballet.*

Maker

Happy Blastoff

Smoke, sound, and fury at the Large
Dangerous Rocket Ship launchpad.

By William Gurstelle

Happy, Texas, (or "The Town Without a Frown" as it is
called locally) is disproportionately well-known considering
it is pretty much just a wide spot in the road: a Texas
Panhandle cow town with a population of 647. The 1999
movie *Happy, Texas*, starring Steve Zahn and William H.
Macy, gave the town a modicum of national attention.
More recently, Happy was the default location for the
weather box on Google's personalized home pages.

Photography by Neil McGilvray and Rockets Magazine

A split second after the main rocket engine ignites, the launchpad is engulfed in smoke and flame.

 In 2006, Happy was important to a large group of makers for another reason — it was the last stop on the Road to Rocketville. Happy was the closest town to the remote launching area where the world's largest conclave of amateur missile makers gathered to show off their biggest and most powerful rockets.

On consecutive July days under the giant sun that bakes the high plains, hundreds of makers converged for the 25th anniversary of the most important event on the calendar of amateur high-power rocket enthusiasts. It's the occasion known as "LDRS." LDRS is an acronym for Large Dangerous Rocket Ships, an appellation whose genesis is apparently lost in the mists of history.

The 2006 LDRS event took place about halfway between the hamlets of Happy and Wayside on a sprawling cattle ranch in sparsely populated Armstrong County, a place with a population density of fewer than 2.4 people per square mile. To get there, fliers drive Texas Ranch Road 287, flat-as-a-three-day-old-Coke, east out of Happy. It's a long, straight, uncrowded chunk of pavement, with only a few power lines, an occasional windmill, a thousand head of live, mooing cattle, and a few cow carcasses scattered about for atmosphere.

The vast emptiness of the long approach makes the appearance of activity and bustle at the launch site jarring. Suddenly, a commotion of technology unfolds. Crowds and tents and noise lend a carnival air. Large, imposing rockets and missiles stand out against the blue sky. People are everywhere, milling and shuffling about, working on their projects, and pointing.

The people at large rocket meets do a lot of pointing — pointing into the sky, arms extended at 70 degrees to the horizon, tracing out a rocket's acceleration skyward and then watching it float down on the end of a parachute or two.

The place is an energy junkie's nirvana: smoke and fire everywhere, rockets roaring up, smoke plumes and contrails hanging like puffy ropes over the ranch, and even the occasional explosion.

These wide open spaces perfectly facilitate the launching and retrieval of the hundreds of rockets that will return to Earth during the course of the event. In particular, this venue has the unusual but highly valued quality that it is well outside all commercial air lanes — the airspace above has no scheduled airplane flights. Even so, organizers must apply for a certificate of special clearance from the Federal Aviation Administration in order to launch the big rockets that attain very high altitudes.

Approaching half a ton, some of these rockets are monsters, the bulk of their weight contained in powerful chemical engines. A single launch can cost more than $800 in rocket fuel.

Catastrophe At Take-Off

The large number of participants keeps several launching pads active. The pads with the biggest rockets are placed the farthest away from people, for occasionally a rocket will blow up, or in rocket lingo, "CATO," on the pad. (Most people here tell me it stands for "Catastrophe At Take-Off.") Nine times out of ten, rocket flights are straightforward affairs, starting with a powerful liftoff, a rocket body zooming vertically to apogee miles up, a perfectly timed parachute deployment, and a soft landing.

But things can go wrong. A design or material flaw in a rocket body can overpressure a rocket casing; then the launchpad gets a shrapnel rainstorm. The crowd is aware of the danger of a rogue rocket coming into the compound on a kamikaze trajectory. Pat Gordzelik, one of the event organizers, takes his turn as range safety officer, or RSO, and becomes the person responsible for the range operations. Whenever a rocket comes in at too high a velocity, the RSO's job is to alert the crowd, which Gordzelik does by shouting "Coming in hot!" over the public address system, with just enough alarm in his voice to make people take notice. Simultaneously he operates an air horn that makes a loud *whoop-whoop-whoop* to make sure everyone gets the message.

Rita Long is among the smallish group of women certified to fly larger rockets. Certification means she's proven herself able to safely build and fly rockets of considerable power. Everyone who flies at LDRS must be certified by a government-recognized rocketry association. Despite that, she says, "The rocket systems inside a rocket sometimes will fail to work correctly. Those rockets, instead of floating down from the sky suspended from a parachute, spiral down to Earth fast and out of control."

Those rockets usually end as "land sharks" or "core samples," i.e., a mangled rocket body and a nosecone buried several feet deep in West Texas caprock. Jokes aside, those situations are dangerous, as rockets with failed reentry systems hurtle down nosecone first and at great speed.

Worse, rockets occasionally flip and turn earthward while the rocket engines are still firing. This makes the situation even more serious, as the rocket is hurtled not only by gravity, but by the chemically induced thrust of the engine as well.

If that occurs with a very large rocket motor, it can be disastrous. Approaching half a ton, some of the rockets launched are monsters, the bulk of their weight contained in their powerful chemical engines. A single launch can cost more than $800 in rocket fuel.

Anatomy of an Amateur Rocket

In and of themselves, rocket engines are marvelous things. The most basic form goes back to first-millennium China, when black powder was first stuffed into bamboo rockets and used to frighten enemy horses. As such, a simple rocket engine is straightforward and easy to understand. Chemical propellant is packed inside a metal casing. The chemicals inside the motor burn and produce hot, expanding gas. The gas rushes out the back of the motor through a nozzle and, by Isaac Newton's third law of motion, the backward gush of the gas results in an equal and opposite forward thrust of the rocket body. Simple, yes. But hey, this is rocket science, and

A. With a whoosh and a roar, a high-power rocket takes flight over the high Texas plains. **B.** New Yorker John Ritz lugs his rocket to the launchpad. **C.** Under pressure, the launch team makes last-minute adjustments.

things get complicated quickly.

Small commercially available model rocket motors consist of black powder propellant pressed under tons of pressure into a hard, dense matrix called grain. When the grain is ignited, it burns linearly, like a cigarette, from back to front. As it does so, it pushes hot gas out through a clay nozzle, and the rocket zips forward until the propellant is all burned up.

The world of high-power rocketry is far more complicated. Instead of simple, black-powder rocket motors, engines are most often made of composite propellant, a rubbery amalgam that consists of an oxidizer chemical such as ammonium perchlorate, and a plasticized binder material that holds the oxidizer in the desired shape and provides fuel.

Composite motors are formed into various shapes, with voids and holes precisely designed to shape the direction and velocity of the exiting gas. Dedicated rocket builders spend days optimizing the contours and figures of their designs and motor formulations. Every rocket maker works with many variables that he or she can adjust to tune the performance: the shape of the rocket body, design of the fins, contour of the nozzle, geometry of the motor's core, mixture of chemicals that make up the propellant, rate of burn, and method of ignition. To become a really good rocket maker takes a lot of scientific, mechanical, and alchemical knowledge. There is also an element of danger associated with toxic and flammable chemicals such as ammonium perchlorate and potassium nitrate.

In the typical large, high-power rocket, the maker builds a finned fiberglass shell that houses the rocket motor, the recovery system, and whatever sensors, cameras, or other payload is carried. (Tripoli Rocketry Association regulations forbid the inclusion of mice, hamsters, frogs, or people as payload.)

The size of a rocket motor is described by a geometrically increasing alphabetic designation. The smallest rockets are called A motors, and provide about 2.5 newton-seconds (N-s) of total impulse, enough energy to power a small rocket to around 500 feet. The B motor is twice as large, having 5N-s of impulse. Because of the doubling scale, some of the big rockets with multiple motors in the L through P range provide enough thrust to win a tug of war with an airliner.

Amarillo Red's Extra Kick

On the last day of the 2006 meet, all eyes turn toward the launch of a massive, multi-stage rocket modeled on the Boeing Corporation's Delta III launch vehicle. Developed by a large group of Kansas-based rocket builders who call themselves the KLOUDBusters, this Delta III is powered by an impressive array of rocket technology: a mammoth P motor, that produces more than a ton of thrust, handmade by Gordzelik and filled with the special ammonium perchlorate-based fuel he developed and named Amarillo Red. The team has encircled the P motor with nine L rockets to provide a super-high-velocity kick at liftoff and again in midair.

Some of the big rockets with multiple motors provide enough thrust to win a tug of war with an airliner.

The smoke, sound, and fury unleashed when the Delta III leaves the ground almost defy description. As the countdown reaches zero, the big P motor and six of the nine boosters roar. In an instant, the rocket clears the launch tower, smoke gushing out in violent eddies, flames spewing from the mouth of the ceramic nozzle. Just 2.8 seconds after liftoff, the three remaining L boosters air-start. A moment later, the rocket is gone from view, only the smoke trail visible in the brightness of the day.

When the booster rockets burn out, they fall away, while the main rocket body continues in ballistic flight. Finally, gravity catches up with it, a full mile up, and points it downward. Inside the rocket body, barometers sense the change in direction and fire the parachute ejection charge. The Delta III floats back down to the wide smiles and bright applause of the rocketeers.

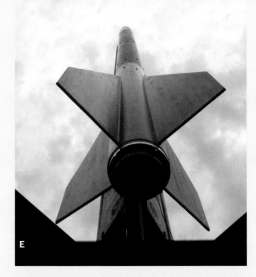

Amateur Rocketry Basics

Model Rockets
Most model rockets use small black-powder motors ranging up to D size. The power and impulse of model rockets are comparatively small. Therefore, safety concerns and costs are minimal.

Mid-Power Rocketry
A step beyond model rocketry are mid-power rocket engines that range from class E to G. While mid-power engines can be powered by black-powder cartridges, very often they're made from composite propellants. Composite engines consist of a rubbery plastic fuel, a powdered oxidizing chemical, and various additives, and they're more powerful on a unit weight basis than black-powder rockets. They generally weigh less than a pound, and can fly much higher than model rockets.

High-Power Rocketry
The largest rockets built with commercially manufactured motors are called high-power rockets. High-power motors range in size from H through O. An O motor is big, roughly 16,000 times the size of an A model rocket engine. Typically, high-power rocket motors require federal licensing and approvals to purchase and fly, and can be flown only at organized club launches held in large, open, unpopulated areas.

Experimental Rocketry
Beyond high-power comes the realm of experimental rocketry. It's for those who choose to build their own rocket motors rather than use commercially manufactured ones. Motors can be any size, though generally they're very large.

Interested in making your own rocket?
Visit these websites for more info:
tripoli.org
nar.org
nakka-rocketry.net
nerorockets.org

D. Composite fuel rocket engine installed and ready for liftoff. **E.** Carefully designed fins keep the rocket on course.

William Gurstelle is a contributing editor for MAKE. His fifth book, *Whoosh Boom Splat*, went on sale in March.

MAKERS VS. SHAKERS

MAKE IT BEAUTIFUL, MAKE IT LAST, OR DON'T MAKE IT AT ALL.

By Saul Griffith

JUST BEFORE THE NEW YEAR, AN EDITOR at MAKE asked contributors for their New Year's resolutions (see makezine.com/go/resolutions). I glibly responded, "to only make things worthy of lasting 100 years." A few months into 2007, I've already broken my resolution, but I've thought more about the reasons behind it, and why I'm still aiming at it.

I was recently in London and visited the British Museum. While standing there in front of a magnificent carved stone piece of the Acropolis, I had to reflect that what we makers are creating isn't particularly impressive. We might be making things, but we are not always being craftsmen — stewards of the materials that have so radically been torn from the earth.

It made me think that the readers of CRAFT magazine (craftzine.com) have the right approach. The average maker is perennially in a state of prototyping. The crafter is making a finished item, lovingly created, designed to last a lifetime or more. It is the difference between spoiled technophile children and Shakers, who built such beautiful furniture that collectors now pay exorbitant prices for simple chairs and tables made 150 years ago.

Why should we care about this distinction? I care because the more prototypes that go to landfill, the worse off the world is. I care because with the loss of craftsmanship, we accept an Ikea world. My father made a teak dinner table for my mother before I was born. More than 30 years later, it's only more beautiful than it was originally. Years of oiling, wine spilling, small hands pawing at it, and countless projects being hammered out on top of it have left it with a loved patina of memory. It would take

a dozen Ikea dining tables to last the same abuse, and that would be a dozen dining tables going to landfill. My father's table will last at least another 70 years with a little love, and a little repair.

So I set myself the task, for this article, to write about something I would make designed specifically to last longer than my own lifetime. I settled on creating for myself a table and benches as functional and beautiful as the table my father built.

The process made me think a lot about electronics, because I couldn't really imagine building anything with a circuit board that could last 100 years, let alone that I'd want to last 100 years. It was a troubling conclusion, and I'm still unresolved regarding the dilemma of making electronic things that will be fun for a month, then fragile and broken for a lifetime.

I had recently built all my office furniture out of bamboo ply, leaving a dozen or so scraps 12"×96" and ¾" thick, so these scraps served as my inspiration and raw materials. At the Squid Labs workshop, with various people looking curiously over my shoulder, I labored over the CAD design for 4–6 hours until the "cartoon" — my colleague Robert "Danny" Daniels' description of CAD — appeared as I wanted it. This was something I didn't want to revise, for to make the "improved version" was to defeat the purpose. Also, to make it in CAD was to leave a digital path that could be followed by others, improved upon, a design I could give away to see perfected by other, brighter minds.

I took my CAD files to the water jet cutter. I could have used more traditional craft techniques, such as pull-saw and chisel artisanry, but my first test attempts showed me that I had neither the patience

This heirloom wouldn't exist without high-tech materials like CAD, a water jet, and a 70-ton press.

nor the skills — another troubling conclusion. The article deadline was looming. I had "real" work to do, and only the weekend to finish this extravagance. Granted, the jet did allow for the incredible accuracy required to make a glueless squeeze fit, but I could already see my deadline-driven world competing with my artisan ideal — or perhaps it's a new artisanry?

I pushed on, I pushed go, the jet started cutting parts, and having already found the machine was designed for something slightly different, I was imagining modifying it to cut wood without getting it wet. The desire for heirloom objects was again coming up against my perpetual prototyping mindset.

It's terrible when the data doesn't support the thesis. Before I had even finished the table, the first of a few design bugs was pointed out to me, and I was contemplating giving it away or trashing it in order to make a "perfect" revision. I was making my heirloom dining table and I'd already noticed mistakes. Nothing terminal, just the things you would do differently next time. "Angle those pieces pointing toward the guests at the center of the table, to prevent sharp corners hitting legs," Danny exclaimed while giving me a lesson in Japanese historical furniture design. I found solace in the fact that Persian rug makers always introduce an error into the rug — only Allah can make things perfectly.

I finished cutting the parts. Now for the assembly: there was the laborious preparation, the sanding, the edge routing and finishing. My impatience built. I was stuck between "finish and look at the mistakes, then revise for the perfect Version Two" and "move slowly, make it perfect now, once, forever, don't waste the material."

As I assembled it, I naturally found all the other things I'd change next time. It all fit together (the CAD modeling had insured that), but it could have been more elegant. Change this, mental note that. If anything, the CAD had made it too accurate. I had to use soap to lubricate the finger joints. Then at the suggestion of Rich Humphry, I realized that a 70-ton press was a better way to put it together than just a rubber mallet — in fact, it was now the only way.

Finally the moment came. I was three-quarters of the way through assembly, and the critical top piece requiring sub-millimeter precision dropped into place. There was my nirvana, my inner peace. I knew at that moment that despite the flaws, it was all going to work. The joy of that moment overcame me. I danced around the workshop as my colleagues looked on. They couldn't understand the revelation I was having. I knew all the flaws in the design, but I understood this object. I had made it. I was going to love it forevermore, precisely and undeniably because I had crafted it. The errors were mine to laugh at and tell stories about at every dinner party to come.

This heirloom was going to be a hoot. I may have failed to mentally resolve the problem of making too many short-lived prototypes and hacks, but at least I succeeded in making one object that my grandchildren might desire.

Thanks to Danny, Rich, Mose O'Griffin, Andrew Forest, and particularly Jim McBride for sanding and providing beer and pertinent derision during the making process.

Saul Griffith works with the power nerds at Squid Labs.

Advice and news for MAKE readers.

■ Have a Reader Service problem that has you ready to bend a circuit?

We aim for perfection, but we admit we sometimes fall short. If you have a problem with a subscription, issues not showing up (ouch), or you can't figure out how to renew (bless you), your first recourse is to contact our customer service department at cs@readerservices. makezine.com, or (866) 289-8847 (U.S. + Canada), or (818) 487-2037 (all other countries). If your problem still hasn't been resolved, please write to me personally at dan@oreilly.com.

■ Find books, kits, tools, apparel, even tickets at store.makezine.com.

Where else are can you find books like *Building the Perfect PC* and *Backyard Ballistics*, downloadable PDF books like *Small Form Factor PCs* and *DIY Coffee*, a complete line of awesome Make-It brand kits like the MAKE Controller and the Open Source MP3 Player, alongside the latest MAKE slogan T-shirts? Only in The Maker Store, of course! Show me that you read this column and enter "dansaid" as your promotion code for a 10% discount on all merchandise bought online at The Maker Store (store.makezine.com) through Aug. 30, 2007.

■ Here's your chance to show your company's support for the maker community.

Hey, we're a small team. We're chugging away on all cylinders to bring this community a couple of awesome magazines, a new line of books, two Maker Faires, free websites, free newsletters, and a ton of support for dozens of fellow groups and associations out there. We're always deeply appreciative of those companies who are willing to step up and help underwrite the effort.

Does your company manufacture products or provide services for the maker community? Want to show your maker colors? Consider purchasing an ad in MAKE magazine or on makezine.com. You can reach a quarter-million makers with a Marketplace ad in MAKE magazine for as little as $500. You can reach hundreds of thousands more makers on makezine.com for as little as $700/week (or $280/week for a text ad).

Interested in showing your support? Contact Katie Dougherty or Dan Woods at advertise@makezine.com.

Maker Faire Takes Gold — Plus Two Maker Faires in 2007!

Maker Faire took "Best Special Event" at the Folio: FAME Awards in March, beating out runners-up Billboard and Rachael Ray! We're seriously blushing.

As this issue hits the newsstand, we'll be having a blast at the 2nd annual Maker Faire held Saturday and Sunday, May 19–20 at the San Mateo County Fairgrounds 20 minutes south of San Francisco. We had over 20,000 makers turn out last year. We may just get twice that many this year — we hope you're joining us.

If you missed Maker Faire Bay Area, consider a fall pilgrimage to Austin, Texas, for the first-ever Southwest Maker Faire — Saturday and Sunday, Oct. 20–21 at the Travis County Fairgrounds. Information about both events can be found at makerfaire.com.

MAKE: Digital Edition Now Open to Web Searches

We've opened up MAKE: Digital Edition to make it searchable and accessible via Google, Yahoo, and other search engines. If you're a subscriber and have opted into the free MAKE: Digital Edition service (and you have your cookies enabled), when you run across a MAKE: Digital Edition link in your web search results, click on the link and you'll have full access to your magazine subscription online. Non-subscribers will have complimentary access to the targeted page plus 3 pages forward and 3 pages backward. Try it out. Let us know what you think.

Dan Woods is associate publisher of MAKE and CRAFT magazines. When he's not working on circulation and marketing or finding cool new stuff for The Maker Store, he likes to hack and build barbecues, smokers, and outdoor grills.

Make: *HOME ELECTRONICS*

All hail the tiny electron!
At the speed of light, these
subatomic servants rush
through circuits, performing
countless tasks to make
our lives more interesting,
comfortable, and fun. Even if
you don't yet know a resistor
from a transistor, the pages
that follow will teach you how
to make electronic devices
that entertain, enlighten,
and enable.

Illustration by Timmy Kucynda

Your Electronics Workbench

WHAT YOU NEED TO GET STARTED IN HOBBY ELECTRONICS.
By Charles Platt

THE BASICS

First, you will need a breadboard. You can, of course, call it a "prototyping board," but this is like calling a battery a "power cell." Search RadioShack online for "breadboard" and you will find more than a dozen products, all of them for electronics hobbyists, and none of them useful for doing anything with bread.

A breadboard is a plastic strip perforated with holes $1/10$" apart, which happens to be the same spacing as the legs on old-style silicon chips — the kind that were endemic in computers before the era of surface-mounted chips with legs so close together only a robot could love them. Fortunately for hobbyists, old-style chips are still in plentiful supply and are simple to play with.

Your breadboard makes this very easy. Behind its holes are copper conductors, arrayed in hidden rows and columns. When you push the wires of components into the holes, the wires engage with the conductors, and the conductors link the components together, with no solder required.

Figure 1 (on page 60) shows a basic breadboard. You insert chips so that their legs straddle the central groove, and you add other components on either side. Figure 1 also shows the bottom of a printed circuit (PC) board that has the same pattern

of copper connectors as the breadboard. First you use the breadboard to make sure everything works, then you transpose the parts to the PC board, pushing their wires through from the top. You immortalize your circuit by soldering the wires to the copper strips.

Soldering, of course, is the tricky part. As always, it pays to get the right tool for the job. I never used to believe this, because I grew up in England, where "making do with less" is somehow seen as a virtue.

When I finally bought a 15-watt pencil-sized soldering iron with a very fine tip (Figure 2), I realized I had spent years punishing myself. You need that very fine-tipped soldering iron, and thin solder to go with it. You also need a loupe — the little magnifier included in Figure 2. A cheap plastic one is quite sufficient. You'll use it to make sure that the solder you apply to the PC board has not run across any of the narrow spaces separating adjacent copper strips, thus creating short circuits.

Short circuits are the #2 cause of frustration when a project that worked perfectly on a breadboard becomes totally uncommunicative on a PC board. The #1 cause of frustration (in my experience, anyway) would be dry joints.

Any soldering guide will tell you to hold two metal parts together while simultaneously applying solder and the tip of the soldering iron. If you can manage this far-fetched anatomical feat, you must

Illustration by Damien Scogin

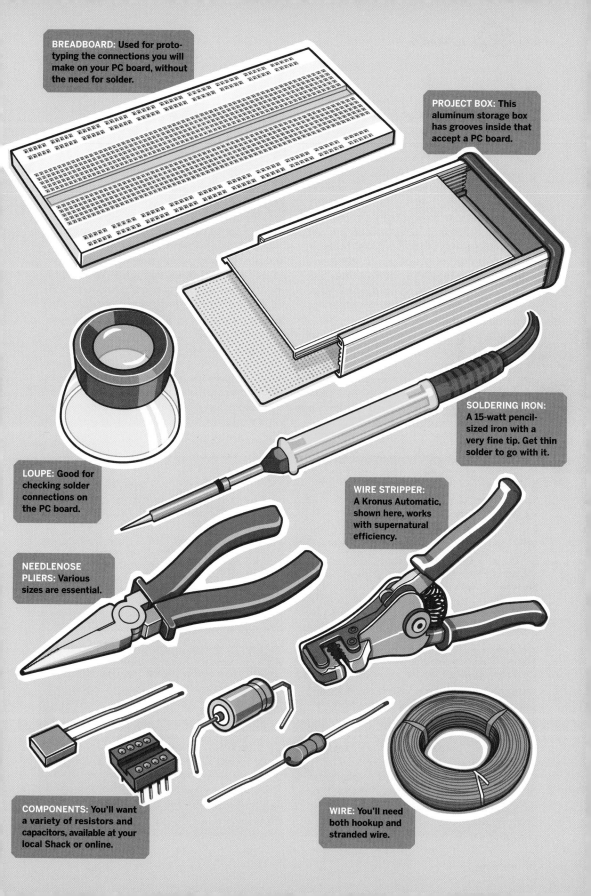

BREADBOARD: Used for proto-typing the connections you will make on your PC board, without the need for solder.

PROJECT BOX: This aluminum storage box has grooves inside that accept a PC board.

SOLDERING IRON: A 15-watt pencil-sized iron with a very fine tip. Get thin solder to go with it.

LOUPE: Good for checking solder connections on the PC board.

WIRE STRIPPER: A Kronus Automatic, shown here, works with supernatural efficiency.

NEEDLENOSE PLIERS: Various sizes are essential.

COMPONENTS: You'll want a variety of resistors and capacitors, available at your local Shack or online.

WIRE: You'll need both hookup and stranded wire.

also watch the solder with supernatural close-up vision. You want the solder to run like a tiny stream that clings to the metal, instead of forming beads that sit on top of the metal. At the precise moment when the solder does this, you remove the soldering iron. The solder solidifies, and the joint is complete.

You get a dry joint if the solder isn't quite hot enough. Its crystalline structure lacks integrity and crumbles under stress. If you have joined two wires, it's easy to test for a dry joint: you can pull them apart quite easily. On a PC board, it's another matter. You can't test a chip by trying to pull it off the board, because the good joints on most of its legs will compensate for any bad joints.

You must use your loupe to check for the bad joints. You may see, for instance, a wire-end perfectly centered in a hole in the PC board, with solder on the wire, solder around the hole, but no solder actually connecting the two. This gap of maybe $\frac{1}{100}$" is quite enough to stop everything from working, but you'll need a good desk lamp and high magnification to see it.

A FEW COMPONENTS AND TOOLS

Just as a kitchen should contain eggs and orange juice, you'll want a variety of resistors and capacitors (Figure 3). Your neighborhood Shack can sell you prepackaged assortments, or you can shop online at mouser.com or eBay.

After you buy the components, you'll need to sort and label them. Some may be marked only with colored bands to indicate their values. With a multimeter (a good one costs maybe $50) you can test the values instead of trying to remember the color-coding system. For storage I like the kind of little plastic boxes that craft stores sell to store beads.

For your breadboard you will need hookup wire. This is available in precut lengths, with insulation already stripped to expose the ends. You'll also need stranded wire to make flexible connections from the PC board to panel-mounted components such as LEDs or switches. To strip the ends of the wire, nothing

Fig. 1: Breadboard (left); upturned PC board.

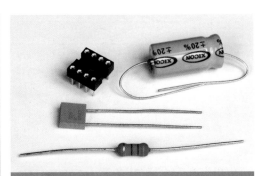

Fig. 3: Socket, big and small capacitors, resistor (front).

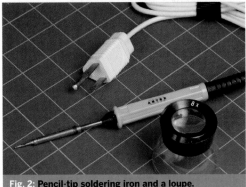

Fig. 2: Pencil-tip soldering iron and a loupe.

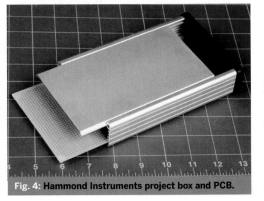

Fig. 4: Hammond Instruments project box and PCB.

Photography by Charles Platt

beats the Kronus Automatic Wire Stripper, which looks like a monster but works with supernatural efficiency, letting you do the job with just one hand.

Needlenose pliers and side cutters of various sizes are essential, with perhaps tweezers, a miniature vise to hold your work, alligator clips, and that wonderfully mysterious stuff, heat-shrink tubing (you will never use electrical tape again). To shrink the heat-shrink tube, you'll apply a Black and Decker heat gun.

If this sounds like a substantial investment, it isn't. A basic workbench should entail no more than a $250 expenditure for tools and parts. Electronics is a much cheaper hobby than more venerable crafts such as woodworking, and since all the components are small, it consumes very little space.

For completed projects you need, naturally enough, project boxes. You can settle for simple plastic containers with screw-on lids, but I prefer something a little fancier. Hammond Instruments makes a lovely brushed aluminum box with a lid that slides out to allow access. Grooves inside the box accept a PC board. My preferred box has a pattern of conductors emulating three breadboards put together (Figure 4). This is big enough for ambitious projects involving multiple chips.

LEARN THE RULES

The final and perhaps most important thing you will need is a basic understanding of what you are doing, so that you will not be a mere slave to instructions, unable to fix anything if the project doesn't work. Read a basic electronics guide to learn the relationships between ohms, amperes, volts, and watts, so that you can do the numbers and avoid burning out a resistor with excessive current or an LED with too much voltage. And follow the rules of troubleshooting:

» LOOK FOR DEAD ZONES. This is easy on a breadboard, where you can include extra LEDs to give a visual indication of whether each section is dead or alive. You can use piezo beepers for this purpose, too. And, of course, you can clip the black wire of your meter to the negative source in your circuit, then touch the red probe (carefully, without shorting anything out!) to points of interest. If you get an intermittent reading when you flex the PC board gently, almost certainly you have a dry joint somewhere, making and breaking contact. More than once I have found that a circuit that works fine on a naked PC board stops working when I mount it in a plastic box, because the process of screwing the board into place flexes it just enough to break a connection.

» CHECK FOR SHORT CIRCUITS. If there's a short, current will prefer to flow through it, and other parts of the circuit will be deprived. They will show much less voltage than they should.

Alternatively you can set your meter to measure amperes and then connect the meter between one side of your power source and the input point on your circuit. A zero reading on the meter may mean that you just blew its internal fuse because a short circuit tried to draw too much current.

» CHECK FOR HEAT-DAMAGED COMPONENTS. This is harder, and it's better to avoid damaging the components in the first place. If you use sockets for your chips, solder the empty socket to the PC board, then plug the chip in after everything cools. When soldering delicate diodes (including LEDs), apply an alligator clip between the soldering iron and the component. The clip absorbs the heat.

Tracing faults in circuits is truly an annoying process. On the upside, when you do manage to put together an array of components that works properly, it usually keeps on working cooperatively, without change or complaint, for many decades — unlike automobiles, lawn mowers, power tools, or, for that matter, people.

To me this is the irresistible aspect of hobby electronics. You end up with something that is more than the sum of its parts — and the magic endures.

Charles Platt is a frequent contributor to MAKE, has been a senior writer for *Wired*, and has written science fiction novels, including *The Silicon Man*.

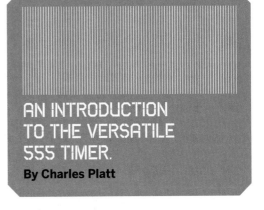

The Biggest Little Chip

AN INTRODUCTION TO THE VERSATILE 555 TIMER.

By Charles Platt

Back in 1970, when barely half a dozen corporate seedlings had taken root in the fertile ground of Silicon Valley, a company named Signetics bought an idea from an engineer named Hans Camenzind. It wasn't a breakthrough concept, just 23 transistors and a bunch of resistors that would function as a programmable timer. The timer would be versatile, stable, and simple, but these virtues paled in comparison with its primary selling point. Using the emerging technology of integrated circuits, Signetics could reproduce the whole thing on a silicon chip.

This entailed some handiwork. Camenzind spent weeks using a drafting table and a specially mounted X-Acto knife to scribe his circuit into a large plastic sheet. Signetics then reduced this image photographically, etched it into tiny wafers, and embedded each wafer in a half-inch rectangle of black plastic with the product number printed on top. Thus, the 555 timer was born.

It turned out to be the most successful chip in history, in both the number of units sold (tens of billions, and counting) and the longevity of its design (unchanged in almost 40 years). The 555 has been used everywhere from toys to spacecraft. It can make lights flash, activate alarm systems, put spaces between beeps, and create the beeps themselves. Today, you can buy a single chip online for about 25 cents.

For the introductory project described below, you can use the 555CN, Fairchild LM555CN or KA555, Texas Instruments NE555P, or STMicroelectronics NE555N. The brand makes no difference. Each manufacturer offers a Complimentary Metal-Oxide Semiconductor (CMOS) version, a dual version, and a surface-mount version in addition to the old-style chip that stands on eight metal legs spaced 1/10" apart. For various reasons, you should use the old-style version.

First I'll show how the 555 can make an LED flash on and off. Then I'll adapt it to generate a musical tone, and finally I'll chain three 555s together to create a gadget you can use to impose a time limit in nonvideo games such as checkers or Scrabble. At the end of a preset interval, the timer will make a groaning sound to tell a tardy competitor that his time's up and his turn is over.

Photography and illustrations by Charles Platt

555 CHIP: The ruler in the background is calibrated in sixteenths of an inch.

1. TURN A 555 CHIP INTO A NOISEMAKER

Figure 1 shows a 555 chip seen from above, with its pins identified. The circular mark stamped in its body is adjacent to Pin 1.

Figure 2 shows a basic light-flashing circuit using the astable mode of the 555, meaning that the output on Pin 3 flips to and fro between positive and negative for as long as the power is switched on. The cycle time is determined by a capacitor and two resistors. A capacitor has electrical storage capacity (hence its name), while resistors reduce the flow of electricity. If you put a resistor in sequence with a capacitor, the resistor slows the charge and discharge times of the capacitor, thus offering a simple way to use electricity to measure time.

When you close switch S1 in the circuit, current flows through R1 and R2 and gradually starts charging capacitor C1. IC1 (the 555 timer) monitors this process. When C1 acquires $2/3$ of the positive voltage powering the circuit, the 555 reverses its output on Pin 3 from positive to negative and forces C1 to discharge itself through R2. When the charge on C1 diminishes from $2/3$ to $1/3$, the chip flips back to its original state, resets its output from negative to positive, and repeats the cycle.

Using a 0.1 microfarad (μF) capacitor for C1, a 120 kilohm (kΩ) resistor for R1, and a 1 megohm (MΩ) resistor for R2, the LED flashes about 5 times each second. (The other components in the circuit have no effect on timing: R3 protects the LED from excessive current, while C2 protects the 555 timer from random electronic noise.)

Suppose you use a 1μF capacitor instead of the 0.1μF capacitor as C1. Now each cycle lasts 10 times as long. Conversely, if you use a 0.01μF capacitor for C1, the cycles are $1/10$ as long. You can also change the timing by adjusting the resistor values. The value of R1+R2 affects the "on" cycle, while R2 alone determines the "off" cycle.

With high resistance and a small capacitor, the 555 will cycle very fast indeed — easily fast enough for its pulses to make musical noises through a loudspeaker.

Figure 3 shows a modified version of the circuit. The LED and its series resistor have been replaced with a different resistor, capacitor C3, and L1, a 1" RadioShack miniature loudspeaker. (Note: You cannot drive a full-size loudspeaker with a 555 timer unless you add an amplifier.) Make sure you update the values of R1, R2, and C1, which have been changed to make the 555 run faster. Now when you connect power, you should hear a low-pitched drone.

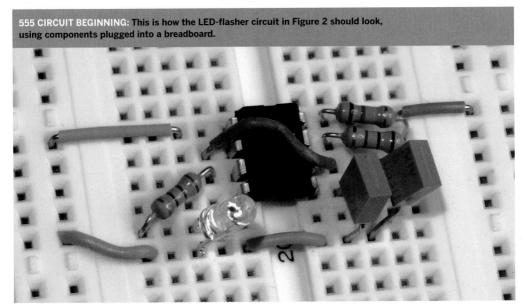

555 CIRCUIT BEGINNING: This is how the LED-flasher circuit in Figure 2 should look, using components plugged into a breadboard.

FIGURE 1

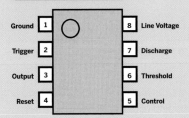

Ground	1	8	Line Voltage
Trigger	2	7	Discharge
Output	3	6	Threshold
Reset	4	5	Control

Pin functions on a 555 timer chip.

FIGURE 2

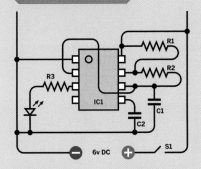

R1: 120kΩ
R2: 1MΩ
R3: 600Ω

C1: 0.1µF
C2: 0.01µF

IC1: 555 Timer

D1: Any LED

S1: Power Switch

FIGURE 3

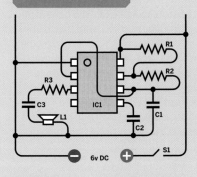

R1: 560kΩ
R2: 560kΩ
R3: 30Ω

C1: 0.01µF
C2: Unchanged
C3: 2.2µF

IC1: Unchanged

L1: 1" Loudspeaker

S1: Unchanged

FIGURE 4

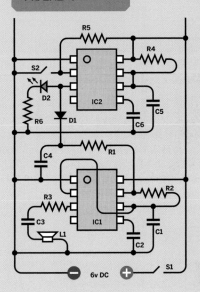

R1 through R3:
Unchanged
R4: 1MΩ
R5: 10kΩ
R6: 600Ω

C1: 2.2µF
C2: Unchanged
C3: Unchanged
C4: 4.7µF
C5: 0.47µF
C6: 0.01µF

S1: Unchanged
S2: Single-pole
momentary
push-button,
normally open.

IC1: Unchanged
IC2: 555 Timer

D1: 1N4148 Signal
diode
D2: Any LED

L1: Unchanged

FIGURE 5

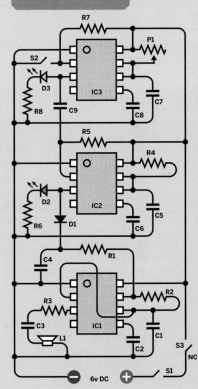

R1 through R6:
Unchanged
R7: 10kΩ
R8: 600Ω

P1: 2MΩ
Potentiometer

C1 through C6:
Unchanged
C7: 2.2µF
C8: 0.01µF
C9: 0.01µF

S1, S2:
Unchanged
S3: Momentary
pushbutton,
normally closed (NC)

IC1, IC2:
Unchanged
IC3: 555 Timer

D1, D2:
Unchanged
D3: Any LED

L1: Unchanged

2. ADD A SECOND 555 TO TRIGGER THE NOISEMAKER FOR A FIXED INTERVAL

We have a noisemaker; now we need to trigger it for a fixed interval. This can be done with a second 555 wired in monostable mode, meaning that it emits only one pulse. Figure 4 shows it added to the circuit. S2 is a pushbutton, although you can improvise just by touching 2 wires together. When this happens, IC2 emits a single pulse lasting about 1 second. This illuminates D2, an LED, to provide visual confirmation. The pulse also goes through D1, a signal diode, and activates IC1, which makes a sound as before, except that C4 prolongs it and causes it to diminish in frequency, creating a groaning effect.

Make sure this version of the circuit works before you continue.

3. ADD A THIRD 555 TO TIME THE WAIT PERIOD

We have a noisemaker that can be triggered for a fixed interval; now we need to measure an interval of time before the sound occurs. A third 555 timer can impose this wait period, if we adjust it with higher-value resistors and a larger capacitor.

In Figure 5, C7 is charged through P1, a potentiometer (variable resistor). You can "tune" P1 to adjust the wait interval, and increase the value of C7 to make the interval even longer. At the end of the interval, the output on Pin 3 goes negative. This connects with the trigger pin of IC2 and tells it to emit its brief pulse, which tells IC1 to make its sound.

Note that S2 has been moved so that it controls IC3. When you use the circuit to impose a time limit during a game, hit S2 at the beginning of each person's turn.

So the circuit won't make its rude noise if a player does move within the allowed time, a cancel-reset button, S3, has been added. You hit this button when a player makes a move. The "NC" beside it tells you that it is a normally closed pushbutton. You still need the power switch, S1, to disconnect your power supply when the gadget is not in use.

555 CIRCUIT COMPLETE: The real-life version of the circuit in Figure 5. The top chip measures a time interval (using a fixed resistor that has been substituted for potentiometer P1). The red LED flashes at the same time that a sound is generated through a 1" loudspeaker (to the left of the chip).

WHAT NEXT?

You can substitute other components instead of the timing resistors to make the 555 behave in interesting ways. In Figure 3, if you use a thermistor or a photoresistor instead of R2, you can control the audio frequency with heat or light. A photoresistor and the 555 in monostable mode can function as a motion detector. Search makezine.com for "555". Also check out doctronics.co.uk/555.htm.

Hans Camenzind never imagined that his timer would become such a universal utility. He now thinks the internal design of the 555 isn't particularly elegant and should have been given a makeover decades ago. Elegance in design can be a big deal for engineers, but for end users, utility is usually more important. The 555 is simple, accurate, and robust, tolerating a wide range of power supplies and able to drive not only LEDs and loudspeakers but also relays and even small motors.

For 25 cents, that's more than enough.

Roomba Hacks

DON'T LET YOUR ROOMBA JUST COLLECT DUST WHEN YOU CAN HACK, MOD, AND TAKE OVER THE WORLD WITH YOUR OWN (CLEANING) ROBOT ARMY.
By Phillip Torrone & Tod E. Kurt

In May of 2006, iRobot, makers of the Roomba robotic vacuum, announced they had shipped more than 2 million cleaning robots, making Roomba one of the (if not *the*) most successful domestic robots in history. With 2 million of anything that can be taken apart, it was only a matter of time before dozens of Roomba hacks hit the net.

Courting this audience, iRobot opened up the interface to all current Roomba models, and released an educational version called the Create. With so many ways to hack these suckers, makers responded by building more projects and developing software. Here's a roundup of some of the interesting projects.

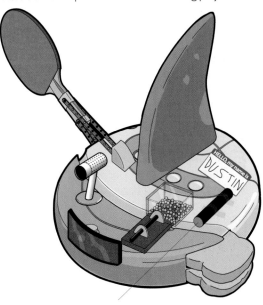

Roomba Costumes
myroombud.com
MyRoomBud was started by a couple of kids who wanted to earn money to buy cowboy boots. They make and design handmade Roomba costumes, including frogs, pigs, tigers, cows, ladybugs, and rabbits. Their motto is, "If you don't dress up your Roomba, it's just a naked vacuum."

As Seen Through the Eyes of the Roomba
roombacam.com
Roomba-Cam is a new site that catalogs videos shot from the point of view of the Roomba. The Roomba kitchen tour puts you in the (autonomous) seat of your vacuuming robot, but nothing is more thrilling than the "nighttime infrared cat hunt."

"The Robot that Vacuums and Serves Web Pages"
makezine.com/go/roombanet
Despite many warnings from science fiction movies and books, Bryan Adams, an MIT Ph.D. student, decided it would be a good idea to use a neural network on a Gumstix Linux board to control a Roomba. Enter RoombaNet. Currently the bot is cleaning Sarah Connor's apartment.

Wii-mote Control
spazout.com/roomba
While Sony and Microsoft have been duking it out over polygons and megahertz, Nintendo brought to market one of the best gaming systems in recent history, the Wii. The motion-sensitive controller makes everything more fun; with Roomba Wii you simply flick your wrist and command your Roomba to do anything. Source code included.

If you have a new Apple MacBook, you can use the tilt sensor in a similar way.

Roomba Music
makezine.com/go/roomidi
The Roomba has a piezo beeper that can play tunes. You've heard it. And its motors make noise. Why not put them under MIDI control? RoombaMidi is Java-based; RoombaMidi2 is written in Objective C and C. Both create a virtual MIDI instrument for use by any Mac OS X MIDI sequencer, like Ableton Live, Logic, and so on.

Cellphone-Controlled Roomba
makezine.com/go/rcontrol
RoombaCtrl is a small Java program for your Bluetooth- and J2ME-compatible phone that works with the build-your-own Bluetooth adapter, as shown in *Hacking Roomba*, or the pre-built RooTooth available at Spark Fun (sparkfun.com).

Cylon Roomba
makezine.com/go/cylon
How does the Cylon base ship keep itself so tidy? With a Cylon Roomba, of course. This tutorial has all you need to make a pulsing LED Cylon Roomba, perfect for cleaning your very own Gaius Baltar or Number Six flat.

Roomba, Get Me a Beer
makezine.com/go/roomba
One of the first robotics projects we all seem to gravitate toward is building a line-following bot. This how-to for modding the Roomba IR sensors shows you how to make your Roomba follow lines on the floor ... maybe even all the way to the fridge.

Bionic Hamster
makezine.com/go/irobot
Using the new iRobot Create programmable robot, a hamster can run around in a clear plastic sphere controlling where the robot goes and how fast it can get there.

Robot Roomba Chimp
makezine.com/go/chimp
We're not living on the moon, or driving flying cars, but we do have an animatronic chimp head that screams atop a Roomba as it cleans.

Tod E. Kurt is the author of *Hacking Roomba*. Phil Torrone is a pioneer of Roomba Frogger and Roomba Cockfighting.

Illustration by Tim Lillis

Nice Dice

BUILD A PAIR OF ELECTRONIC RANDOM NUMBER GENERATORS. By Charles Platt

For me, a good construction project should create an object that is fun, functional, and pleasing to the eye — and if it teaches me something interesting along the way, so much the better. I managed to satisfy all these requirements when I designed and built a pair of electronic dice. Although dice simulations have been around for many years, I was able to simplify the project while at the same time making it more interesting.

The basic principle is easy to understand. One chip generates a rapid stream of pulses. A second chip counts them and displays each number via an array of LEDs imitating the pattern of spots on a die. At an arbitrary moment, we stop the process to display one number at random.

Most dice circuits use a decade counter to do the counting, but I went for a 74LS92 chip, which counts in sixes and returns the result in binary code. This may sound uninviting, but it just happens to be ideal for a dice display.

Figure 1 (page 71) shows the 74LS92 seen from above. It has 14 pins, of which Pins 9, 11, and 12 are outputs. On each of these pins a higher voltage represents a 1 while a lower voltage represents a 0. The chip starts with 0 on all three pins.

When it receives a timing pulse, Pin 12 counts from 0 to 1. At the next pulse, Pin 12 carries the 1 over to Pin 11 and resets itself to 0. Then it counts to 1 again, and then, since Pins 11 and 12 have both reached their maximum, they carry a 1 over to Pin 9 and reset themselves to 0. The binary counting sequence is

shown in the table in Figure 2.

If Pin 12 drives an LED in the center of the array, and Pin 11 is wired to two diagonally positioned LEDs, and Pin 9 adds the LEDs in the opposite corners, the counter will display all the spot patterns from 1 through 5 in correct numerical order. I liked the simplicity of this arrangement, but there was one little problem: when the counter begins with 0 on all three pins, it should really display a 6. To fix this, I needed a NOR gate.

Computers function with Boolean logic, meaning that they contain many components that take two or more inputs and give an output according to a simple rule. The rule for a NOR gate is that when all of its inputs are low, it gives a positive output.

A chip known as the 74LS27 contains three 3-input NORs (as shown in Figure 3). All I had to do was connect all three outputs from the counter to the three inputs on a NOR gate, and then wire the NOR output to light up six LEDs. Figure 2 shows this symbolically.

The actual circuit is in Figure 4, with components laid out as you are likely to place them on a bread-

Photography and illustrations by Charles Platt

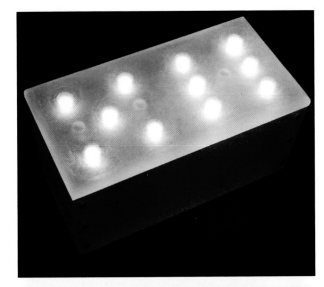

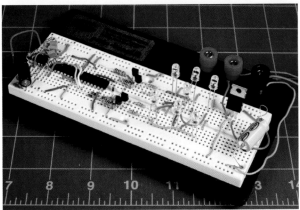

board. For the sake of clarity, I omitted the wires supplying positive and negative voltage. You'll have to link all the plus (red) symbols and all the minus (blue) symbols with the positive and negative sides of a power supply, which must provide a properly regulated 5V DC. The best way to guarantee this is by passing a higher voltage (for instance, from a 12V AC adapter) through a regulator such as the Fairchild LM7805CT. I've always liked this component since I read the manufacturer's spec sheet, which describes it as "essentially indestructible." Now, that's a semiconductor I can live with.

I had to add four diodes around the 74LS27 chip to prevent outputs conflicting with inputs. I needed transistors to drive the LEDs, because the counter chip doesn't have enough power to do this on its own. Resistors R5 through R8 prevent the transistors from overloading the chip, and R9 through R12 protect the LEDs. Their exact values will depend on which LEDs you happen to use. Check the manufacturer's recommended voltage (usually somewhere between 2.5V and 3.5V), apply a meter to each LED while it's illuminated, and adjust the resistor values up or down accordingly.

Most likely you will mount your LEDs on a separate display panel, in a configuration such as the one in Figure 5. The little dice icons in the LED circuit indicate inputs that correspond with outputs from the circuit in Figure 4. Just connect a wire between each matching pair of icons.

When you have finished building your circuit, you should short the unused pins on the 74LS27 chip together and ground them, to reduce the chance of errors caused by electronic noise.

S1 at the top of the circuit is a momentary pushbutton, supplying power to a 555 timer, which sends about 500 pulses per second to the counter. The LEDs become a blur, and no one can tell which number is being generated at any moment.

When the button is released, this is equivalent to throwing the die. Power to the 555 is cut off, but capacitor C2 has accumulated a potential and now slowly discharges itself. As its voltage diminishes, the 555 runs more slowly, until the LEDs finally stop flickering and show one number — much like a die as it rolls across a table and stops with one face up.

If you want two dice instead of one, you will need to duplicate the entire circuit (except for the 74LS27 chip, which still has two spare NOR gates on it). To ensure randomicity, the second die must run at a different speed from the first, requiring slightly different values for C3, R2, and R3. You can double the value of C2 so that the second die takes longer to stop flashing, like the second reel on a Las Vegas slot machine. The two dice displays should still start simultaneously, which will entail using a double-pole pushbutton for S1, providing power separately to each circuit.

If you need to troubleshoot the circuit, try adding a 10µF capacitor in parallel with C3 to make the 555 timer run very slowly. Disconnect IC3 and use a meter to check voltage on the outputs from IC2. If they are counting properly, reconnect IC3 and check its outputs one at a time. Be sure to use a power supply of only 5V; any more will burn out the counter chip.

Of course I could have simulated dice more easily by writing a few lines of software to generate random numbers on a screen, but even a fancy screen image cannot have the same appeal as a well-made piece of hardware. Also, I derived satisfaction from using simple, dedicated chips that demonstrate the binary arithmetic and Boolean logic that are fundamental in every computer. Best of all, I ended

PARTS LIST

S1: **SPST momentary pushbutton switch**

R1: **100Ω resistor**
R2: **100K for time increment**
R3: **100K for time increment**
R4: **100Ω**
R5, R6, R7, R8: **10K for IC protection**
R9, R10, R11, R12: **500Ω for LED protection (adjust the values to suit your LEDs)**

C1: **100µF capacitor for decoupling**
C2: **22µF for timer slowdown**
C3: **0.01µF for time increment**
C4: **0.01µF for decoupling**

D1, D2, D3, D4: **1N4148 (or similar) signal diodes**
Q1, Q2, Q3, Q4: **BC550 signal transistors**
IC1: **NE555N (or similar) timer**
IC2: **NTE 74LS92N (or similar) counter**
IC3: **74LS27 triple 3-input NOR gates**

Also:
7 LEDs **(your choice)**
LM7805CT **(or similar) 5V voltage regulator**

Source for chips: ebay.com **or** mouser.com

7400s on the Moon:

If you noticed the similarity between part numbers for the 74LS92 counter and the 74LS27 NOR gate, there is a reason for this. They both belong to the pioneering 7400 family of integrated circuits developed by Texas Instruments back in the day. The 7400s travelled on NASA's manned moon missions and have become legendary in electronics, even meriting a Wikipedia entry (wikipedia.org/wiki/7400). 7400s are still used for prototyping and for teaching computer science. It doesn't take much Boolean logic to combine AND, OR, NOT, NAND, NOR, XOR, and XNOR gates in ways that will take you far beyond electronic dice.

up with an object representing my particular tastes and idiosyncracies; and to me, that's what making things is all about.

Charles Platt, a frequent contributor to MAKE, has been a senior writer for *Wired*, and has written science fiction novels, including *The Silicon Man*.

FIGURE 1

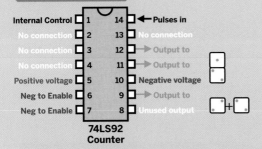

Internal Control	1	14	← Pulses in
No connection	2	13	No connection
No connection	3	12	→ Output to
No connection	4	11	→ Output to
Positive voltage	5	10	Negative voltage
Neg to Enable	6	9	→ Output to
Neg to Enable	7	8	Unused output

**74LS92
Counter**

The 74LS92 counter requires power on Pins 5 (positive) and 10 (negative) and negative voltage on Pin 7 to enable it. Pins 9, 11, and 12 can be connected through transistors to LEDs representing spots on a die, to display spot combinations 1 through 5; negative=0, positive=1.

"No connection" means that the pin is not connected to anything inside the chip and can be ignored.

FIGURE 2

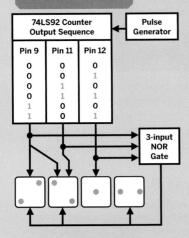

74LS92 Counter Output Sequence			Pulse Generator
Pin 9	**Pin 11**	**Pin 12**	
0	0	0	
0	0	1	
0	1	0	
0	1	1	
1	0	0	
1	0	1	

3-input NOR Gate

A NOR gate creates a spot pattern representing number 6 if, and only if, it sees a 0 on all three outputs from the counter.

FIGURE 3

Input 1A →	1	14	Positive voltage
Input 1B →	2	13	← Input 1C
Input 2A →	3	12	→ Output 1
Input 2B →	4	11	← Input 3A
Input 2C →	5	10	← Input 3B
Output 2 ←	6	9	← Input 3C
Negative voltage	7	8	→ Output 3

**74LS27
Triple 3-input
NOR Gates**

This chip contains three NOR gates with three inputs each. All inputs for each NOR gate are functionally identical.

The output is positive if, and only if, all three inputs are negative (negative=0, positive=1).

FIGURE 4

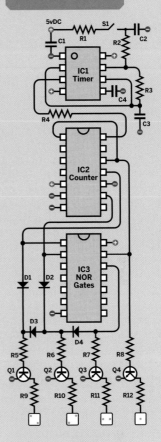

Each dice icon indicates an output that should be connected to the LED display.

FIGURE 5

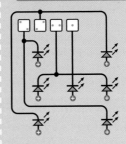

Seven LEDs are arranged in this pattern to emulate the spots on a die. The dice icons indicate inputs that correspond with outputs from the circuit in Figure 4.

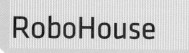

RoboHouse

EASY-TO-IMPLEMENT HOME AUTOMATION.

By Andrew Turner

My house is a robot. It thinks, reacts, predicts, and informs. Throughout the day it lets me know how its inhabitants are doing and takes care of all the little things I forget. If I'm worried that I left the front door open or that the heater is turned up too high, I can view my house's website or RSS feed through a browser or my mobile phone.

In addition to making my life easier, my house is concerned about saving the planet and my wallet. It can turn off unused appliances and lamps and intelligently control my heating and air conditioning system according to when someone is home and where it's coldest or hottest in the house.

The power my house wields is easy to implement. In fact, I rent my house, so all of my additions, modifications, and alterations are temporary. They can be pulled out tomorrow, and my house would revert to its boring, uninformed structure.

The central brain of the house is an Apple PowerMac running Perceptive Automation's Indigo (perceptiveautomation.com). Indigo is a very powerful and configurable home automation application that allows me to register all my devices, sensors, and scripts. Once I input the devices in my house, I can control them using the provided user interface, or I can create customizable, dynamic "Control Pages" using my Mac's web server. These web pages can be a layout of my house, security camera and status, virtual

appliance interfaces, or whatever I want to create using the layout tools in Indigo. The Control Pages use a mixture of HTML and JavaScript to create a very easy-to-use interface that's accessible from anywhere in my house, or even at the office or when traveling, using my laptop or mobile phone.

For controlling lamps and appliances, I use a mixture of X10 (x10.com) and Insteon (insteon. net) control modules, available from places like smarthome.com or machomestore.com. These modules plug into any 2- or 3-prong wall outlet and have a unique serial number that Indigo uses to address and control.

I also use a mix of AppleScript, Ruby, and Python scripts for additional intelligence. These scripts are fired off when I come home or leave, to do computer backups, start playing music through my Squeezebox/SlimServer, or pull down traffic information and display it on my Nabaztag, a wi-fi-enabled, programmable, electronic rabbit (nabaztag.com).

Photography by Greg Ruffing

A TYPICAL DAY AT MY AUTOMATED HOUSE

7:00 a.m.: WAKE UP
» Alarm wakes me, and bedside lamp slowly brightens if it's still dark outside.
» Coffee maker begins to brew.
» Computer wakes up, downloads email and news, and displays calendar.
» Stereo turns on and begins to play music.

8:00 a.m.: TIME TO LEAVE FOR THE OFFICE?
» Nabaztag gives me the traffic conditions, alerting me to when I should leave to make any appointments.
» Coffee maker shuts off.
» Stereo shuts off.
» Lights left on are turned off.

12:00 p.m.: WHAT'S GOING ON AT HOME?
From work, I view the status of the house.
» Did that package arrive?
» Are the pets sleeping?
» Is the house secure?

2:00 p.m.: IS ANYONE HOME? NO
» Start vacuuming.
» Check MythTV to record any TV shows.

5:00 p.m.: I'M HEADING HOME!
When I pull into the driveway, the house swings into action.
» Entry lights turn on.
» Computer wakes up.
» Stereo begins to play.

7:00 p.m.: TIME FOR DINNER AND A MOVIE
» Pull up recipe book in kitchen.
» When I sit down to eat, check MythTV for shows.
» Dim the lights.

10:00 p.m.: BEDTIME
» Computer reminds me that I should get to sleep for an early appointment in the morning.
» Once I'm in bed, the lights all turn off.
» Computer backs itself up before going to sleep.

THE
AUTOMATED
HOUSE
OF TODAY

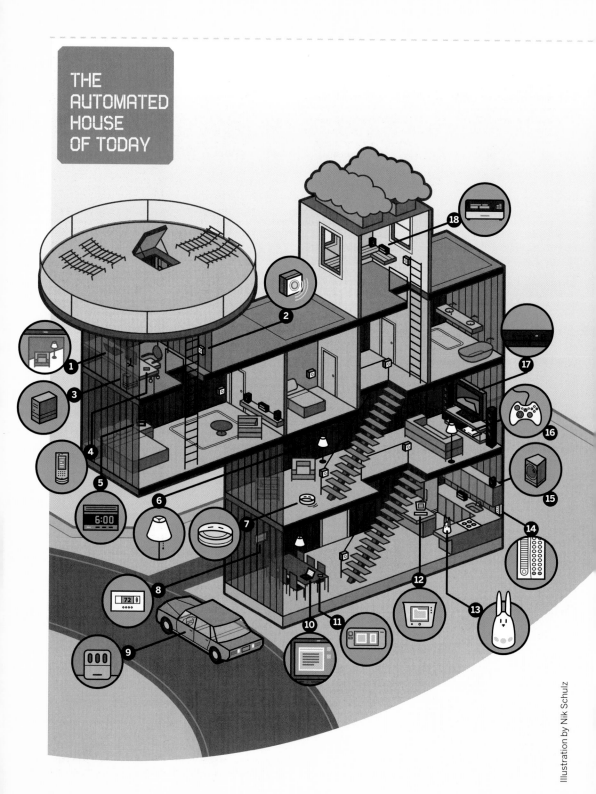

Illustration by Nik Schulz

✚ You can find more information on the scripts, hardware, and configuration at my automation site: automation. highearthorbit.com

1 iSight webcam lets me check out portions of the house.

2 Based on my route through the house as seen by motion detectors, the lights turn on or off, and the computer knows if I'm going to my office or to the kitchen.

3 The computer grabs email, news, calendar. Records TV on MythTV. Data is backed up each night.

4 Mobile phone Bluetooth locks/unlocks the computer when I approach my office and starts/stops the music.

5 Alarm clock goes off and tells house to wake up. In the evening, I am able to turn off the entire house from here.

6 Lamps controlled by automation system and by various "scenes" in the house (movie, wake-up, etc.).

7 Roomba vacuums if no one is home.

8 Temperature sensors feed the HVAC system to turn on the heater or air conditioner or fans.

Author Andrew Turner in his office, where he makes the automated magic happen.

9 The car has HomeLink to notify the house that I'm arriving or leaving.

10 Laptop and **11** Nokia N800 allow web-based interface to house.

12 3Com Audrey allows me to pull up recipes, activate music, and control the house from the kitchen.

13 Nabaztag lets me know what the traffic is like, and whether my wife or I am heading home.

14 Wireless X10 PalmPad and switches provide simple interface to lights, devices, and scenes.

15 Music is streamed around the house using Slim Devices Squeezeboxes.

16 An Xbox is running Xbox Media Center (XBMC) for streaming MythTV, music, videos, photos, etc.

17 All appliances can be turned on and off using Smarthome's wall socket ApplianceLincs.

18 Slim Devices Squeezebox.

An aerospace engineer, Andrew Turner is a freelance developer of geospatial technologies. He enjoys photography, neural networks, and curling (the kind on ice).

Propeller Chip

BASIC STAMP'S CHIP GRACEY PUTS A NEW SPIN ON MICROCONTROLLERS.

By Dale Dougherty

The head guy at Parallax, Chip Gracey, is truly self-taught, which means that he has had to find his own way. Twenty years after teaching himself to program the first generation of personal computers, the creator of the new Propeller microcontroller still speaks with the enthusiasm and amazement of a bright teenager: "The tools are out there. These days with the internet, it is so easy; you can learn anything. What used to be obscure stuff that only a few people were interested in — well, today those people put it on the net to share among themselves, and the rest of us have access to it."

Lately he's been "playing around" with speech synthesis. "I've put about two steady years of work into it. Just now, I'm on the cusp of having a working voice synthesizer." He immediately starts walking down the idiosyncratic path he took to get where he is today, and I do my best to follow along. He conveys all the details as though they are keys to finding the levels of an adventure game he created.

He explains how the key to speech synthesis is to reproduce vowel sounds, which are vocal resonators. The long sounds, the *oohs* and *ahhs*, resonate inside the hollow spaces in our skulls when we say words like "food" or "bath." If he's able to replicate vocal resonance in software, the result will be something much closer to normal speech.

"I used to look at the human voice on a scope in high school and you can see the waveforms of speech. But you can't look at speech in terms of waveforms because speech is really a spectral phenomenon." So he believed the key to his approach was to write programs that allowed him to visualize speech patterns in more detail. "If you can visualize it accurately then you know what you have to do to re-create it."

"I kept reading about digital resonators. Everybody talked about them but no one explained how they work. I couldn't find a recipe. Then I finally came upon CORDIC math, which means coordinate rotation. [CORDIC is an acronym for Coordinate Rotation Digital Computer.] It was developed in the 50s. Real interesting stuff and very simple."

He points to the computer screen where there are lines of code that don't explain themselves. "I realized that a CORDIC rotation algorithm could be used to create a resonator. You just add in the stimulus vector (add x's and y's) and then rotate the

Photography by Robyn Twomey

resulting point. Repeat. Repeat. Repeat. Any in-band energy will have the effect of growing the distance of the point from (0,0) as it rotates at the resonant frequency. This is key to making vowel sounds."

I probably look more than a little confused. He adds: "You've got to excite that resonator. It's like a bell. *Bing-gggg*. That took me forever to figure out but it was very simple." I'm not sure I follow everything he says, but I get that here is a guy who works and works an idea and pushes it forward, bit by bit, and eventually gets somewhere. He works at a level of electronics that is completely ethereal to most of us, but not to him.

"Ultimately, everything has to be physical so we can perceive it with our senses," Gracey says. "The hardware is like the body and the software is like the spirit. The whole point of the Propeller chip is to make an able body for the kind of spirits you want to create. Like speech synthesis."

Gracey's immediate goal is to write a state-of-the-art sound synthesis library for the new Propeller chip. He explains that existing speech synthesis applications "almost always require a dedicated speech chip." By adding sound synthesis as a software capability with the Propeller, he can make it affordable to create applications that generate a vocal track. He adds: "The whole point is to enable inventors to make stuff they couldn't make before."

From the beginning, Parallax has offered an embedded platform, distinct from the PC and Windows, for hobbyist and educational projects, not to mention professional uses. Why not just use a PC?

Gracey says a PC can get expensive for a number of applications, especially when compared to using the BASIC Stamp controller. However, he says the "big sleeper issue is reliability because Windows seems to blow itself up in time."

Neither BASIC Stamp nor the Propeller have an operating system. Why? "The systems are so simple that the code you write is the operating system." He sees the microcontrollers as deterministic systems that do what you tell them to do; put an operating system onboard and you will have something with a mind of its own. PCs are interrupt-driven, which means the operating system is designed to interrupt what it's doing if it thinks it should be doing something else; a PC usually hosts multiple applications. Embedded applications typically are designed to do one thing — such as read an RFID tag and unlock a door, or monitor and store GPS coordinates of a rising weather balloon — very well.

The landmark BASIC Stamp controller is a custom circuit based around an 8-bit Microchip PIC chip, running Parallax's now famous BASIC interpreter, plus a high-level development environment. Other models in the BASIC Stamp family are based on a Scenix SX processor or may come with more I/O

pins, faster speeds, or other enhanced features.

The new Propeller chip has eight separate 32-bit processors called "cogs" that can independently process information, with a shared memory space. The Propeller is an entirely custom integrated circuit design of Gracey's own making, and a completely new development environment and programming language (called Spin). It's a completely new path that he set out on, and he's got to convince customers to go along with him.

On the Parallax user forums, (forums.parallax. com) one user wrote that the new Propeller chip was untested and unproven and that he'd only consider using it in a hobby application. Gracey wrote back:

"I know that the Propeller is solid because I designed, debugged, tuned, and tested it myself. The only other person involved in the silicon design was one other layout engineer. This took eight years of my time, and two years of the layout engineer's time. An excruciating amount of attention went into every aspect of the Propeller's design and testing, and I allowed no compromises."

As Gracey sees it, he's always been ahead of his customers. Ever since he learned to program, he knew he was doing something that other people found valuable. "I would think of something that might be neat and I'd make it, and then I could sell it to people."

Today, Parallax employs 40 people, most of them located in an industrial park in Rocklin, Calif., about a half-hour from Sacramento. None of them are salespeople. "We have nobody calling anybody trying to make them buy anything," Gracey says proudly.

Gracey's business career started in his bedroom when he was in high school. He created a software utility for disk duplication on the Commodore 64. In 1987, he made $130,000 in royalties; he was only 17. Next he made development tools for the Apple II. He was so into his programming that he barely finished high school. He tried a year of college to please his

mother, but quickly dropped out.

Then something happened to him and he dropped out of business as well. Today, he calls it a "malaise," a term Jimmy Carter used to characterize America's "crisis of confidence" in 1979. Gracey says that he thought the reason he liked to program was to buy cool stuff. He spent some of his money on a stereo, a car, remote control toys, video games, and the like.

Even though at 19 he was still living with his parents, he had everything he could ever want, but he discovered none of it made him happy. He got to the highest level of his own adventure game, only to find out he didn't care anymore about the achievement. Game over.

He recalls: "You're trained to think that you do things for money, that success equals money." Gracey was all mixed up and it took him a while to separate the work he enjoyed from the rewards it had gained for him. "Money is not the object. What I liked about what I was doing, was doing it," he said. "What was making me happy was the learning and the creativity."

So he came back and started Parallax with a friend, Lance Walley, and they eventually shared an apartment where they also worked. They began making microcontroller development tools for the Intel 8051, an all-in-one computer chip, meaning it had a CPU, RAM, ROM, and I/O on board. (Intel's 8086, the heart of the first IBM PC motherboard, is just a CPU.) Why the 8051? "I thought it would be neat to build development tools for a single chip."

Then in 1993 he developed the BASIC Stamp product. "I loved all these little microcontrollers because they are so much fun. The trouble was that you had all this arcane setup to do to program them. I wondered: what if you could make a little computer that you could program in a high-level language and that was about as cheap as a microcontroller?"

He saw that Microchip had PIC chips, a popular family of 8-bit microprocessors notably used in Microsoft PS/2 mice, and 8-pin EEPROMs (Electronically Erasable Programmable Read-Only

> ## " MONEY IS NOT THE OBJECT. WHAT I LIKED ABOUT WHAT I WAS DOING, WAS DOING IT. WHAT WAS MAKING ME HAPPY WAS THE LEARNING AND THE CREATIVITY. "

Memory), simple serial-based devices that store small amounts of data. He realized he could develop a BASIC interpreter that he could program into the Microchip PIC device, and then use the EEPROM to hold the user-generated code that the PIC would pull out and execute. In addition, he created a development tool on the PC that's a high-level compiler. "You write code on the PC, which compiles it down and then you hit a button and send the code down to the chip and it starts running."

That's the essence of the BASIC Stamp line of products, the bread-and-butter of Parallax, which made it easier for more people to program microcontrollers. For several years, Gracey and Walley were getting orders, as many as 20 a day, and assembling the circuit boards and kits themselves, in their apartment. "We worked really hard," recalls Gracey. Now Parallax has machines that automate the assembly of the circuit boards.

Walley and Gracey parted ways in 1996, and Chip's younger brother, Ken, came on in 1997 to help manage operations as the company began to grow. I wondered if Ken, who had finished college, was considered the good brother in the family. "For a long time, my parents really felt like their hopes

had been lost in me," Chip recalls. "Now they think of us both as good brothers."

An environmental studies major, Ken has picked up quite a bit of the technical side of the business as well as handling day-to-day operations. "He's not as enthused about the technical side the way that I am, but he can always learn as much as he needs to know." Ken is the multitasking manager handling sundry activities and interfacing with customers while Chip remains singularly focused on development. Parallax has grown up to become a satisfying family business, with their father, Chuck, coming in most days to work as well.

The decision to build the Propeller chip was driven not only by Gracey's developing interest in chip design. (On his bookshelf, he points out *Principles of CMOS VLSI Design* by Weste and Eshraghian, which he says taught him all he needed to know.) He felt that large chip manufacturers were not willing to experiment with new design ideas. Moreover, he felt the chip manufacturers had too much control in the relationship. "We had to have our own silicon," he says. He had to go his own way.

There's a rock-star-sized poster of the Propeller chip's schematic on the wall in Gracey's office, and he gives me a tour of all its functional areas, which he knows so well. "The Propeller has eight capable processors and a shared memory. Any one of those can synthesize speech, generate a VGA signal, talk to a mouse or a keyboard, or digitize something."

He's not the first to put multiple processors on a chip. However, previous approaches required parallel processing to use the chip; the challenge is to figure out how to split up a program so that parts of it can run in parallel. "That's not what you do with multiple processors," says Gracey defiantly. "Instead you let them each do something different."

The chief design goal of the Propeller was to allow multiple processes to run concurrently without interrupting each other. For the new speech synthesis library, this means each processor can produce a voice independently of the others. "I've added vibrato,"

EXCITE THAT RESONATOR: Gracey's spectrograph program shows a recording of "1 2 3 4 MAKE magazine" followed by some whistling.

Joe Grand, a member of MAKE's Technical Advisory Board, has worked closely with Parallax and has followed the development of the Propeller.

Chip Gracey designed the Propeller chip (parallax.com/propeller) completely from the ground up, a feat rarely attempted in a world where most products are based on some existing prior art, reference design, or chipset.

Even the most fundamental base of the Propeller, having multiple "cog" processors, each active for a given time "slice" and sharing a common memory space, is sufficiently unique and different from existing microprocessor technologies.

I remember visiting Parallax for the first time a few years ago; Chip had just got his code working to output graphics to a TV monitor using an FPGA development system — the precursor used to test code before moving to actual silicon.

He showed me the code in his excited fashion, and we spent the next few hours hacking away at different parts of it, trying to display multiple Parallax logos on the screen. His enthusiasm was unparalleled and contagious. I knew he was on to something big. To see the fruits of his labor come together at an early, pre-release Propeller training session was just phenomenal.

Even with the release of the Propeller, the BASIC Stamp family is still a hugely popular device for the hobbyist and electronics community, and I personally don't see that going away. Much to the chagrin of Chip, who I think wants

BIRTHPLACE OF CHAMPIONS: A Parallax solder stencil for the now-classic Basic Stamp II.

to move everyone over to the Propeller, the two product lines seem to attract different audiences (at least for now).

On one hand, using Stamps for simple, low-cost tasks is just so easy. I absolutely love using them for quick prototypes that aren't worth the effort or energy to deal with a more complicated microprocessor development environment. On the other, I've never seen such a unique and interesting device as the Propeller, and the Spin code people have written for it already is amazing, even in its very short life.

Gracey says. "You can have all eight voices singing, just like a choir." (Months later, he sent me a cool sample called "Propeller Monks" generated by his code; hear it at makezine.com/10/propeller.)

It took years to develop the Propeller chip. The path was not straight, and there were many obstacles. It required new machines, new investment, and new knowledge of manufacturing techniques. It also took longer than Ken hoped it would. Perhaps recalling something that had worked so well before Chip had his midlife crisis at 19, Ken promised to buy him a very high-end stereo system, an audiophile's dream, if he could complete the design on time. And that's where we stop in an otherwise empty

office, and sit down, listening to a jazz standard on the turntable. "You can hear so much more with this system," says Gracey, with the glee of a kid having friends over to listen to his record collection.

One thing Gracey said earlier in the day stands out. "If you have a sense of mastery, even if it's not complete, then it goes a long way toward helping you get to where you want to go." It's a statement as much about himself as about what Parallax hopes its products do: help others to find their own way.

Dale Dougherty is editor and publisher of MAKE.

THE MACHINE ROOM

We built up our own lab that gave us the "hands" and "eyes" that we needed to work on our chip. First, we invested in a Micrion FIB (focused ion beam) machine (shown at right) that allowed us to perform microscopic surgery, so that we could check failure hypotheses and make experimental modifications. Think "wire cutters," "soldering iron," and "solder" for the sub-micrometer wiring inside the chip.

The other big thing we acquired was a Schlumberger e-beam prober — essentially a scanning electron microscope that can use its electron beam to measure voltages on those same tiny wires while the chip runs at full speed. Think "7GHz, non-loading, 10nm-tip, contactless oscilloscope."

These machines are almost *Star Trek* in their technology, and they get you all the way down to where you need to be in order to see and fix problems.

We were able to purchase these machines, used, for only 0.5% of what they cost new. The real investment, though, turned out to be the six months it took to get them running, and to learn how to use them.

Now, we can even do our own maintenance on them, which is not trivial. All this was a huge adventure in itself, but invaluable in getting the Propeller's silicon perfected.

—*Chip Gracey*

TURBO SALAD: Chip offers this backstory: "This is a turbo molecular pump [used to pump stray molecules of air from the vacuum chamber where chips are fabricated] that crashed. It weighs about 45 pounds and has about 10 pounds of stacked turbine blades on ceramic bearings that spin at 40,000 rpm. The bearings wore down critically over a weekend and the pump seized within a few revolutions, resulting in 'turbo salad.' A tremendous amount of kinetic energy was let loose and the 9/16" stainless steel bolts were all bent 10 degrees in a circular pattern as a result of the sudden breaking. We usually get our turbo pumps rebuilt when they are starting to show wear through either excess whine or noticeable heat. Getting such a pump rebuilt is about $1,500. Buying a rebuilt used one costs about $4,000. Buying a new one can cost over $20,000. So, our tactic is to rebuild, and if we blow that, get a used rebuilt one."

Turbine photograph by Dale Dougherty

The Spin Zone

A WHOLE SYSTEM OF POSSIBILITIES – EVEN AN ENTIRE COMPUTER – AWAITS YOU IN THIS SIMPLE CHIP.

By Ken Gracey

ABOVE: Gaming guru Andre LaMothe's Hydra Game Console (hydraconsole.com) is the first commercial application based on the Propeller chip.
BELOW: The Propeller Proto Board can be used for permanent projects.

The Propeller's architect, my brother Chip, has roots in vintage computing, which helps explain the "whole system" features designed into this single chip. The chip performs basic microcontroller functions very easily, yet it can also be an entire small computer once you add the keyboard, mouse, and display.

The Propeller can execute 160 million instructions per second across its 8 processors while consuming only 80mA. Each processor can be dedicated toward a single task, and variables are globally available to other processors in Propeller. There are no interrupts in this chip — the concept is that you'd dedicate a cog to managing repetitive, externally driven tasks. A configurable timer makes scheduled events straightforward.

This hardware is made useful through an object-oriented, high-level programming language called Spin. The Spin interpreter is actually built into the Propeller's ROM hardware. Spin looks a bit like BASIC and C++, but customized for the Propeller's architecture.

Propeller programmers can go to the Parallax Propeller Object Exchange and download hundreds of code modules for serial communication, VGA/TV display, mice, math, keyboards, motor control, and sensors, and link them together in a "top" Spin program. Beginners can use objects off-the-shelf, and engineers can customize. You can also program the Propeller in good ol' assembly language.

AND WITH THIS POWER?

Applications possible with a single Propeller include:
» CNC controller. Simultaneous generation of stepper motor signals, receiving serial or parallel data from a PC, user interface with a mouse and VGA/TV display, and interface for E-stop buttons and limit switches.

» Oscilloscope and signal generator. Electronic data interpretation, adaptable triggering.
» Sensor processing and robotic navigation. Concurrent processing of data from GPS, ultrasonic, infrared, compass, and encoder sensors while generating PWM for motor control.
» Video games. Simultaneous interface to joystick, keyboard, and mouse, with Propeller-to-Propeller networking, all while displaying the game on TV.

But it's the simple projects where most of us are already putting Propeller to use. Parallax's history in educational documentation and personalized support should give you some comfort in your first experience.

➕ Go to parallax.com/propeller to get started programming the Propeller. There, you'll find kits, educational labs, and free downloads of all object code, plus a manual and the Propeller software tool.

Ken Gracey is vice president of Parallax, Inc.

Shots from a Revolution

"Desperation and insecurity." That's what Mark Richards says drove him to photograph the machines in *Core Memory: A Visual Survey of Vintage Computers* (text by John Alderman, Chronicle Books, May 2007). After a long freelance career shooting hot spots like Afghanistan in the 1980s for *Time* and *Newsweek*, Richards realized that photojournalism was dying, and went on a frantic search for something new.

He found it on a visit to the Computer History Museum in San Jose, Calif., where he was struck by the unintended beauty of early computers built in military-funded university labs. Using a digital camera wired to a laptop monitor, Richards' complex lighting schemes and 30-second exposure times show the rest of us what he first saw in his mind's eye.

—*Mark Frauenfelder*

ABOVE: Philco 212, 1962. Memory: 64K. $1.8M.   BELOW: ENIAC, 1946. Memory: Ten 10-digit numbers. $500k.

Burroughs ILLIAC IV, 1975. The fastest computer in the world at the time cost $31 million and ran until 1982.

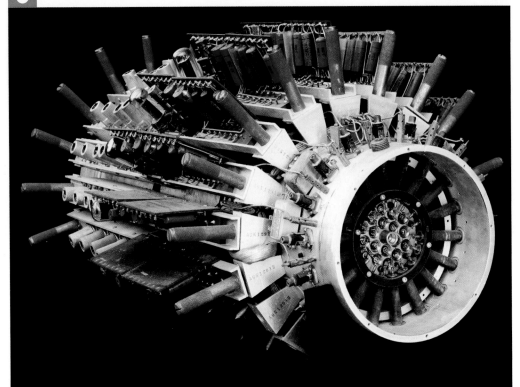

UNIVAC I, 1951. The first commercial computer in the United States boasted 20K of mercury delay-line memory.

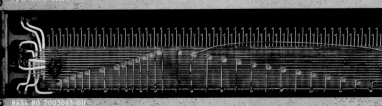

S/N RAY P-001

NASA NO 2003H4-011 X
MFD BY RAYTHEON CO
S/N RAY 36

ERASABLE DRIVER MODULE B9-10

NASA NO 2003D04
REV C1
MFD BY RAYTHEON CO
S/N RAY. P001

ERASABLE DRIVER MODULE B9-10

ENGINEERING PROTOTYPE

NASA NO 2003026-011
REV A1
MFD BY RAYTHEON CO
S/N RAY P001

CURRENT SWITCH MODULE B11

ENGINEERING PROTOTYPE

NASA NO 2003043-011
REV A 2
MFD BY RAYTHEON CO
S/N RAY P001

SENSE AMPLIFIER MODULE B13-14

ENGINEERING PROTOTYPE

NASA NO 2003043-011
REV 7
MFD BY RAYTHEON CO
S/N RAY P004

SENSE AMPLIFIER MODULE B13-14

ENGINEERING PROTOTYPE

NASA NO 2003027-011
REV D1
MFD BY RAYTHEON CO
S/N RAY P001

STRAND SELECT MODULE B15

ENGINEERING PROTOTYPE

Apollo Guidance Computer, 1965. Each NASA moon mission used two of these $250k computers with 4K RAM.

Make: Projects

Deep-cycle your mind with a sound-and-light show at brain wave frequencies. More interested in a no-brainer? Bend a mod desk set out of pliable ABS plastic and you'll have your hands on a truly miraculous material. Or, if you prefer getting your hands dirty, take pointers from the scientists who worked on Biosphere II and engineer an ecosystem with your own Tabletop Shrimp Support Module.

The Brain Machine

88

ABS Plastic Desk Set

100

Biosphere in a Jar

110

THE BRAIN MACHINE

By Mitch Altman

Photography by Sam Murphy

RIDE YOUR OWN BRAIN WAVES

You don't have to be a Buddhist monk to meditate, or a Sleeping Beauty to sleep well. Achieve these altered states of consciousness, and others, with this simple microcontroller device.

What would happen if you could play a recording of brain waves into someone's brain? That question popped into my mind one day while I was meditating, and it turns out that there are devices that can do this. Sound and Light Machines (SLMs) produce sound and light pulses at brain wave frequencies, which help people sleep, wake up, meditate, or experience whatever state of consciousness the machine is programmed for. The first time I tried one was a trip! Not only did I follow the sequence into a deep meditation and then out again (feeling wonderful!), but along the way I had beautiful, outrageous hallucinations.

This article shows you how to build an SLM for much cheaper than you can buy one. We'll do it the easy way, by hacking a microcontroller project that already exists: Limor Fried's Mini-POV kit. This cool toy blinks pictures and words in the space you wave it through, and we can transform it into an SLM simply by changing the firmware and some minor hardware.

Set up: p.92   Make it: p.93   Use it: p.99

Best known for inventing TV-B-Gone, a keychain that turns off TVs in public places, **Mitch Altman** is interested in any technology that gives people more choices for improving their lives.

HOW SOUND AND LIGHT HACK THE BRAIN

The brain produces varying proportions of brain wave types, depending on its current levels of relaxation, focus, and other mental states. Each type of wave has its own characteristic frequency range, which can be read by electroencephalography. Many people's brain waves will synchronize to lights and sounds pulsing at brain wave frequencies, and this makes the brain change its state — a process called "entrainment." By playing sequences of pulses into your eyes and ears, you can program your brain to follow any brain wave experience you like.

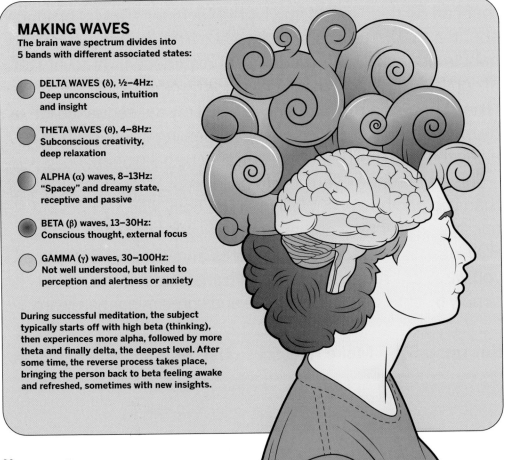

MAKING WAVES

The brain wave spectrum divides into 5 bands with different associated states:

- **DELTA WAVES (δ), ½–4Hz:** Deep unconscious, intuition and insight

- **THETA WAVES (θ), 4–8Hz:** Subconscious creativity, deep relaxation

- **ALPHA (α) waves, 8–13Hz:** "Spacey" and dreamy state, receptive and passive

- **BETA (β) waves, 13–30Hz:** Conscious thought, external focus

- **GAMMA (γ) waves, 30–100Hz:** Not well understood, but linked to perception and alertness or anxiety

During successful meditation, the subject typically starts off with high beta (thinking), then experiences more alpha, followed by more theta and finally delta, the deepest level. After some time, the reverse process takes place, bringing the person back to beta feeling awake and refreshed, sometimes with new insights.

Illustrations by Timmy Kucynda

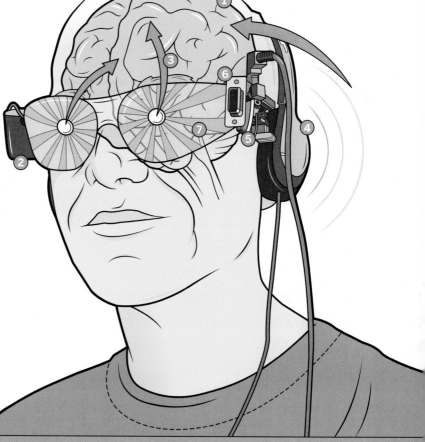

HOME ENTRAINMENT SYSTEM

We'll program our SLM to follow a 14-minute sequence that tracks the meditation experience. Since the device generates only one frequency at a time, it phases in new brain states by switching frequencies back and forth. For example, to go from fully awake to somewhat dreamy, we generate beta for a while, then alpha, then toggle between beta and alpha, reducing the duration of beta and increasing that of alpha with each iteration. Our code's brainwaveTab array defines the full sequence.

1. Brain entrains to the programmed wave sequence, and alters its state accordingly.
2. Battery pack powers the electronics.
3. LEDs in front of user's (closed) eyes pulse light at 2.2, 6.0, 11.1, or 14.4Hz, in order to elicit delta, theta, alpha, or beta waves, respectively.
4. Headphone speakers play different tones into right and left ears, to produce binaural beats (see below) that match the LED pulse frequencies.
5. Microcontroller on circuit board runs the firmware, the program that resides in the microcontroller, which controls the LEDs and headphones.
6. Serial port connector writes the firmware into the microcontroller, letting you program your own brain wave frequency sequences.
7. Graphics simply look cool.

BINAURAL BEATS

Instead of simply playing the entraining wave frequency through both headphone speakers, we employ a more effective method. When we play different frequencies into each ear, the brain perceives a binaural beat frequency just as if the two tones were played next to each other on guitar strings. The beat results from the two tones cyclically reinforcing and canceling each other out, at a rate that equals the difference between the frequencies.

To generate a beta binaural beat, we play a 400Hz tone in one ear and a 414.4Hz tone in the other. The user perceives a sound, sort of like "wah-oo-wah-oo-wah," that fades in and out 14.4 times per second.

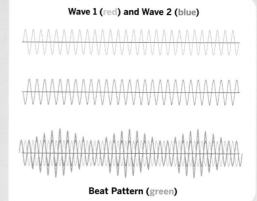

Wave 1 (red) and Wave 2 (blue)

Beat Pattern (green)

SET UP.

MATERIALS

[A] Safety glasses **from your favorite hardware store**

[B] Headphones **Cheap ones, like the free ones on airplanes**

[C] MiniPOV v3 kit **$18 from** makezine.com/go/povkit. **SLM firmware downloadable with full documentation from** makezine.com/10/brainwave

[D] 1kΩ resistors, ¼W **(2) Mouser Electronics (**mouser.com**) part #660-CF1/4C102J**

[E] 1.0μF capacitors, bipolar **(2) Mouser #647-UVP1H010MDD**

[F] 3.5mm stereo jack **that mates with your head-phones, Mouser #161-3402**

[G] AA batteries (2)

[H] Kynar wire **Also known as 30-gauge wire-wrap wire, any 2 colors; I used blue and yellow, RadioShack #278-502 and #278-503.**

[I] Silicone adhesive **Also known as RTV (room temperature vulcanizing); clear or any other color, about $5 from your favorite hardware store**

[J] Toothpicks **for spread-ing silicone adhesive**

[K] Cable ties, 4" long x 0.1" wide (6) **Mouser #517-41932**

[L] 4" length of ¹⁄₁₆" heat-shrink tubing **Mouser #5174-11161**

[M] Permanent or dry-erase marker

[N] 3M Magic Tape

Eye graphics **(optional) Download and print the template I used at** makezine.com/10/brainwave.

Rubbing alcohol and toilet paper **for removing perma-nent marker marks, if you make a mistake**

Total materials cost should be $23–$56, depending on what you can scrounge.

TOOLS

Computer **with 9-pin serial port or a USB port and a USB/serial adapter based on the PL2303-chipset, about $15 for port**

Soldering equipment and solder

Needlenose pliers

Diagonal cutters

Wire strippers

Drill and ³⁄₁₆" drill bit

Small, sharp nail

"Third hand" tool **(optional) Very handy**

Scissors, hobby knife, printer **(optional) for graphics**

MAKE IT.

BUILD YOUR SOUND AND LIGHT MACHINE

1. BUILD AND TEST-PROGRAM THE PCB

The core of our electronics is the MiniPOV v3 kit. We'll be referring to the kit's website at ladyada.net/ make/minipov3.

1a. Follow the excellent instructions on the MiniPOV website to solder all the components to the included printed circuit board (PCB) *except* the following: LED1, LED2, LED3, LED4, R5, R6. Go ahead and leave these 6 components out.

NOTE: In electronics-speak, soldering a component onto a PCB is called "stuffing" it.

1b. Insert the 2 AA batteries into the battery holder and switch on. The microcontroller will run the MiniPOV firmware, and the 4 stuffed LEDs should light up.

1c. If the LEDs don't light up, then debug. Are the batteries in correctly? Check the power connections. Check for bad solder connections or solder bridges (shorts between 2 pads). Check that all LEDs, and D1, D2, and D3, are not stuffed backwards. Now we'll change the MiniPOV's firmware so you can learn how to use the development tools and see how easy it is.

1d. Download and install the **AVRDUDE** software needed for your operating system. Links for the software are on the MiniPOV website listed above, under "Download." If you're using a USB/serial adapter, also download the USB/serial converter driver.

1e. Create a folder on your computer called *slm*. Download the firmware for MiniPOV from the MiniPOV website into the *slm* folder and unzip it. Using a text editor, open up the *mypov.c* file, and change the pattern in the image array (near the top of the file) to the pattern at right. Since a 1 will light up an LED and a 0 leaves it off, this will create a pattern on 4 LEDs that looks something like "VVVVV" when you wave the MiniPOV back and forth through the air. Switch off the MiniPOV's battery pack, plug the MiniPOV into your computer's serial port, and switch it back on.

```
B8(10000000)
B8(01000000)
B8(00100000)
B8(00010000)
B8(00100000)
B8(01000000)
```

1f. Compile and program the microcontroller by following the instructions on the MiniPOV website for your operating system. Under Windows, you'll enter the following (at right) into a command window. Turn the MiniPOV power pack off and unplug the PCB from the serial port. Switch the power back on and the 4 LEDs will light up. Wave it around, and behold the VVVVV pattern you just programmed in!

```
> cd slm
> del mypov.hex
> make mypov.hex
> make program-mypov
```

NOTE: make mypov.hex **compiles the code into a hex file that the microprocessor can run, and** make program-mypov **uploads the hex through the serial port to the microprocessor.**

2. MAKE THE SLM CONTROLLER

2a. Download *SLMfirmware.zip* from makezine.com/10/brainwave into your *slm* directory and unzip it. Let it overwrite the makefile with the new one if asked. Program the microcontroller with the new firmware. Follow Step 1e above to hook it up to your computer. Then compile and upload the firmware to the micro. The process is identical to Step 1f, but we're programming *slm* instead of `mypov`. Under Windows, enter the code seen to the right.

```
> cd slm
> del slm.hex
> make slm.hex
> make program-slm
```

2b. Turn off and unplug the MiniPOV, which is now an SLM controller, although it won't do much, since we don't have any outputs hooked up yet. Solder the two 1µF capacitors into the pads for LED3 and LED4. Solder the two 1kΩ resistors into the pads for R5 and R6. Clip the excess leads on the back of the PCB.

NOTE: Adding the capacitors and resistors creates low-pass filters that smooth out the square waves into sine waves, for more pleasing audio.

2c. Cut a 2" length of blue wire for hooking up the stereo headphone jack (for ground). Cut two 2" lengths of yellow wire. Strip ⅛" off one end of each wire, and tin.

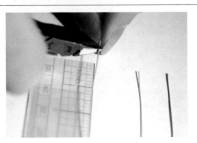

NOTE: To "tin" a wire means to heat it up with the soldering iron and melt a little solder on it. This makes it easier to solder it to something else.

2d. Place the stripped end of the blue ground wire into the headphone jack's ground terminal and solder. Place one of the yellow speaker wires into one of the jack's empty terminals and solder. Do the same for the remaining speaker wire.

NOTE: There are 3 terminals on the headphone jack. One is ground, offset from the others, and the other 2 are for the left and right speakers. For this project it doesn't matter which way we connect the left and right speakers.

2e. Twist these 3 wires neatly, but leave about ½" untwisted. Cut them so that they all extend the same length, strip ⅟32" off each, and tin. Solder the ground wire (blue) to the PCB trace that connects all of the LED grounds together. Then solder the right and left speakers (yellow) to the outputs for LED3 and LED4.

2f. Test the audio by plugging your headphones into the jack, putting them on, and flipping the power switch. You should hear some spacey sounds in both ears. If not, the most likely problems are bad solder connections or solder bridges.

3. PREPARE THE BATTERY PACK

3a. Unsolder the 2 battery pack wires from the PCB.

3b. Cut two 4" lengths of wire (1 yellow, 1 blue), strip ⅛" off 1 side of each, and tin the stripped ends.

3c. Hold the blue wire to the black battery pack wire (negative) and solder them together. Slip ½" of heat-shrink tubing over the connection and lightly rub the tubing back and forth on all sides with the soldering iron for a few seconds until the tubing shrinks around the connection. Repeat for the red battery pack wire (positive), but use yellow wire to extend it.

NOTE: Although you can use electrical tape to cover the exposed wires, heat-shrink looks nicer and won't unravel.

4. PREPARE THE GLASSES

4a. The glasses need 1 LED facing each eye. Place the glasses on your face, look straight ahead, and point the tip of a marker at your right eye. Slowly move the marker toward your eye to make a mark directly in front of your eye. Repeat for left eye. If you want to cover the glasses with cool graphics, trace the shape of each lens onto a piece of paper, and cut the 2 templates out for later.

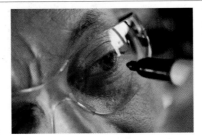

NOTE: The marks on the glasses should be somewhat symmetrical. If not, use some toilet paper and rubbing alcohol to remove the marks and try again.

4b. Use a small sharp nail to make a pilot indentation into the 2 marks on the glasses. Drill a ³⁄₁₆" hole for each eye, taking care not to let the bit slip. Test-fit the leftover LEDs from the MiniPOV kit into the holes. If you can't push them all the way in, enlarge the holes a little by pushing the sides against the spinning drill bit.

4c. Push an LED into each eye hole so that they will face your eyes when you wear the glasses. Cut four 8" lengths of wire (2 yellow, 2 blue), strip ⅛" off 1 side of each, and tin the stripped ends.

4d. The longer lead of each LED is the positive lead. Cut off all but ⅛" of the positive leads for each LED, tin them, and solder them to 8" yellow wires. Do the same for the negative leads, soldering them to the blue wires.

NOTE: Cut the leads one at a time, to make sure you don't confuse the positive and the negative.

4e. Bend the wires on both LEDs up and over the top edge of the glasses and glue each LED into place from the outside using silicone adhesive. Use a toothpick to create a solid hemisphere of silicone that makes good contact with the glasses and coats the LED leads and exposed wire. Let the adhesive harden (1–2 hours).

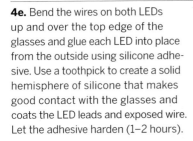
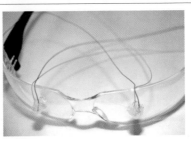

NOTE: Silicone adhesive is a good choice for electronics projects because it does not conduct electricity and is easy to use.

5. ATTACH THE BATTERY PACK AND PCB

5a. Prepare 2 pairs of cable ties by slipping one about ¾" through the other.

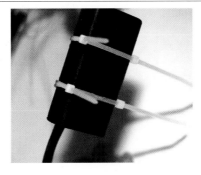

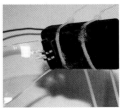

NOTE: Using 2 short cable ties instead of 1 long one makes it cleaner to go around corners.

5b. Spread silicone adhesive 2⅝" along the right temple of the glasses to span the length of the battery pack. Press the battery pack into the adhesive and secure with the double cable ties. Trim the excess length off the cable ties and then let the adhesive harden.

5c. Route the 4 wires from the LEDs and 2 from the battery pack along the top of the glasses toward the left temple (where they will connect to the PCB), twisting them together along the way. Keep the 2 wires from the battery pack together, so you don't confuse them with the LED wires.

5d. Hold the PCB in place on the left temple (but don't glue yet), positioning it such that the connector sticks in front of the left lens when the glasses are unfolded. This will let you plug the glasses into your computer. Trim the 6 wires so that there will be only a little bit of slack when they connect to the LEDs and battery pack. Strip ¹⁄₁₆" off the end of each of the 6 wires, and solder them to the unstuffed LED and battery pack contacts, maintaining proper polarity.

NOTE: Test the LEDs by turning on the battery pack power switch. The LEDs on the glasses should flash. If not, then double-check the wiring.

5e. Spread silicone adhesive 2¼" along the left temple of the glasses for the length of the PCB. Press the PCB into the adhesive and secure with another cable tie pair. Remember that the connector needs to plug into your computer's serial port without the glasses getting in the way. Clip any excess wire under the PCB and cut the excess off the cable ties.

6. SECURE THE HEADPHONE JACK AND LOOSE WIRES

6a. Push the wires to the side, put a blob of silicone over D2 and R11 on the PCB, and push the headphone jack sideways into the blob. Add more adhesive with a toothpick to completely cover the terminals of the jack and about the first ⅛" of each attached wire.

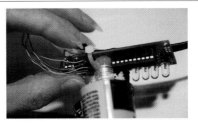

NOTE: Gluing the wires and terminals will prevent shorts and provide strain relief.

6b. To hold the loose wires in place while they are being glued, prepare eight 4" strips of Magic Tape by folding a small bit of each end over to stick onto itself. Stick the tape pieces where you can reach them while gluing the wires to the glasses in the next step.

NOTE: Doubling the ends over makes it much easier to pull off the tape strips after the adhesive hardens.

6c. Use a toothpick to thinly spread silicone adhesive between the wires and along the top of the glasses, then glue the wires along the glasses. Use the tape strips to keep the wires in place while the adhesive hardens. Also tape the headphone jack in place. Wait for the adhesive to harden, about 1–2 hours, then remove the tape.

7. APPLY COOL GRAPHICS (OPTIONAL)

7a. Using the templates you created earlier, cut out lens-shaped pieces from your graphics for the left and right lenses. Cut a hole in each one where the back of the LEDs will poke through.

7b. Apply a very thin layer of silicone adhesive (spread with your finger) in a horizontal strip on the back of each cutout. Affix them to the front of the glasses. Now they look cool!

FINISH ⊠

NOW GO USE IT »

TRIP THE LIGHT FANTASTIC

Get comfortable, put on the glasses and headphones, close your eyes (the LEDs are bright!), and flick the power switch. Enjoy the hallucinations as you drift into deep meditation, ponder your inner world, and then come out after the 14-minute program feeling fabulous.

New to meditating? Here are a few ideas. Breathe normally and pay attention to your breathing. If you space out, no worries — just focus on your breathing again. Or don't focus on anything; follow your thoughts wherever they go. You can also concentrate on something before you start: solving a technical problem, exploring a difficult decision, feeling through the issues in a relationship. If you space out, just refocus on your intent. There is no end to the variety of what you can do in a meditative state.

> **⚠ CAUTIONS:** Blinking lights should be avoided by anyone prone to seizures. If you are sensitive to seizures, please disconnect the LEDs in this project; your brain can effectively entrain with the binaural audio beat alone. Certain conditions, such as ADHD, can worsen with theta and delta entrainment. Please discontinue using this SLM if you experience any problems.

PROGRAMMING YOUR OWN BRAIN WAVE SEQUENCES

You can program your own sequence by editing the brainwaveTab definition in the file *slm.c*. Each line defines a pair (bwType, bwDuration) that sets the brain wave type for an amount of time, specified in ten-thousandths of a second. For example, the first pair, { 'b', 600000 }, runs the beta frequency (14.4Hz) for 60 seconds.

You can program the session to last longer than 14 minutes, spending more time in deeper states. To aid sleep, you may also want to decrease the pitch of the audio, since lower pitches are more relaxing. You can do this by tweaking the OCR0A setting in main and the OCR1A settings in the function do_brainwave_element, also in the file *slm.c*. Explore and enjoy!

BRAIN MACHINE II

I'm working on a more complex SLM with many improvements to enhance relaxation and entrainment. It will gently fade sound and lights in and out, handle multiple simultaneous frequencies, match brain wave types to different colored LEDs and binaural base frequencies, and mask the binaural beats with music or other sounds. Physically, it will fit in a soft fabric sleep mask and use a light, flexible, rechargeable battery. It will have other enhancements as well; I'll keep the MAKE community posted.

RESOURCES

Megabrain Power: New Tools and Techniques for Brain Growth and Mind Expansion by Michael Hutchison, Ballantine Books, 1996

The High-Performance Mind by Anna Wise, Tarcher, 1997

Dreamachine Plans by Brion Gysin, Temple Press, 2006

The Anna Wise Center: annawise.com

The Monroe Institute: monroeinstitute.com

AVR Freaks (microcontroller forum): avrfreaks.net

➕ More resources at makezine.com/10/brainwave

PLASTIC FANTASTIC DESK SET

By Charles Platt

ABS TO THE RESCUE

Last night I dreamed of a magical material that would be bendable like metal, as easy to shape as wood, and would never warp, split, or splinter. It would be washable, would never need painting, and would last almost forever.

This morning, when I sat down at my desk, the stuff from my dream was right in front of me. It fact, it had been there for several months, ever since I made a pen rack from ABS.

ABS is *acrylonitrile butadiene styrene*, a plastic that really does have dreamlike qualities. If you've ever picked up a Lego block, you've handled ABS. Car stereo installers and model railroad buffs sometimes use it, but craftspeople and hobbyists generally have been slow to adopt it. You can saw it, drill it, sand it, whittle it, and drive screws into it, and it never warps, splits, or splinters. Best of all, you can bend it quickly into complex shapes by using a simple gadget that costs around $200.

To acquaint you with its pleasures (and a few quirks), I'll describe how to build a page stand — a simple work aid that facilitates copy-typing by holding pages upright beside your video monitor. After that I'll describe a portable CD caddy and a pen rack, and will suggest more projects you can make.

Set up: p.103   **Make it:** p.104   **Use it:** p.108

Charles Platt is a frequent contributor to MAKE, has been a senior writer for *Wired*, and has written science fiction novels, including *The Silicon Man*.

Photograph by Sam Murphy

"JUST ONE WORD: PLASTICS"

There's a great future in it.

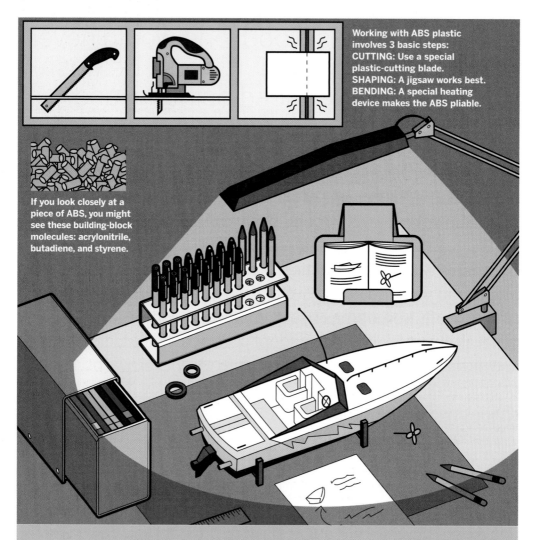

Working with ABS plastic involves 3 basic steps:
CUTTING: Use a special plastic-cutting blade.
SHAPING: A jigsaw works best.
BENDING: A special heating device makes the ABS pliable.

If you look closely at a piece of ABS, you might see these building-block molecules: acrylonitrile, butadiene, and styrene.

ABS: THE WONDER MATERIAL OF THE MILLENNIUM

Since ABS doesn't rot or rust, whatever you build is likely to be around for a very long time. I imagine people in the future shaking their heads in amazement. "Look at this," they'll say. "Those guys in 2007 used to build almost anything out of plastic, back before oil cost $10,000 a barrel." Actually, plastic can become a renewable resource, if it is made from biofuels. Manufacturing it will always require more energy than using wood, though, so let's use it sparingly to build appropriate, good-looking things that are designed to last — while we can still afford to do so.

Illustration by Nik Schulz

SET UP.

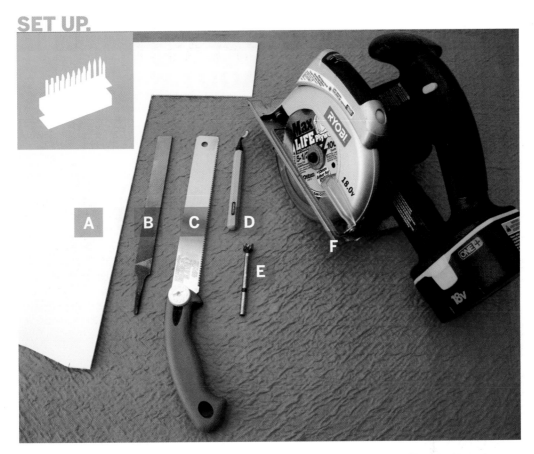

Pieces of ABS a couple of feet square are available online, but you'll save money if you truck on down to your nearest plastic supply house and buy it like plywood, in 4'×8' sheets. To discover whether you actually have a nearby plastic supply house, search for "plastic supply" in your yellow pages or Google Local. HobbyLinc.com has ABS sheet and a lot of extruded structural shapes for model making. eStreetPlastics has limited stock at good prices at stores.ebay.com/eStreetPlastics. Regal Piedmont Plastics, online at piedmontplastics.com, has a huge inventory and many supply centers around the nation, but you'll have to collect it yourself, and they may not be willing to cut 4'×8' sheets into smaller pieces.

Stock colors include black, white, and "natural," which is beige. Sheets usually are textured on one side, which is the side that should face outward, since it is more scratch-resistant than the smooth side. To build the page stand, you will need white ABS, ⅛" thick (see picture sample at right).

MATERIALS

[A] White ABS plastic sheet ⅛" thick

[B] Coarse metal file (optional)

[C] Japanese-style, pull-to-cut saw **I use the Vaughan Extra-Fine Cross-Cut BearSaw, 9½", 17 tpi. Check your local home improvement store.**

[D] Deburring tool

[E] Drill with ½" Forstner bit

[F] Handheld circular saw **Ryobi 18V, 5½" blade (from Lowe's or mail order)**

Plastic bender **Available from FTM Inc. (thefabricatorssource. com). Some other clever plastic-working tools are available here, too. (not shown)**

ALTERNATIVE TOOLS
Ultimate jigsaw **DeWalt DC330 XRP with Bosch T101BF blades (available from Lowe's or mail order)**

Table saw and blade **Freud LU94M010, 10", 80-tooth industrial plastics blade**

NOTE: ABS often comes in sheets that are textured on one side, smooth on the other. The textured side withstands more abuse.

MAKE IT.

DESKTOP PAPER CADDY

START ⋙   Time: **1½ Hours**   Complexity: **Easy**

1. PREPARE

Because the plastic surface you end up with is the surface you start with, you'll have to be careful not to scuff or scratch it while working. Clean your bench thoroughly before you begin, taking special care to remove any metal particles, which will tend to become embedded in the plastic. Use wooden shims in the jaws of your vise, and avoid resting the plastic accidentally on any sharp tools or screws. Working with ABS requires a clean environment and a very gentle touch.

2. MAKE PRELIMINARY CUTS

You'll bend up on the blue dashed line, and down on the red dashed lines. To make rounded inner corners, drill ½" holes (blue circles) before cutting. There's nothing else to it: no fasteners or additional sections. Very often an ABS object can be fabricated by making multiple bends in a single piece.

Begin by cutting a 7"×16" rectangle out of a larger piece. Because ABS is not brittle, you cannot just drag a knife to score a groove and then snap it, as you can with acrylics. A saw is necessary. The bad news is that if you use a table saw, the plastic will tend to melt and stick to the blade. This will lead quickly to kickback, in which the blade grabs your workpiece and hurls it at you powerfully enough to break bones. If you have extensive experience using a table saw, you are actually more vulnerable, because the reflexes and cautions you have developed while dealing with wood will not be adequate for working with soft plastic. Please take this warning seriously!

 WARNING: Avoid using a table saw to cut plastic. If you must use one, install a special plastic-cutting blade.

Fortunately there is a simple answer: buy a plastic-cutting blade, which has a larger number of thicker teeth to absorb the heat.

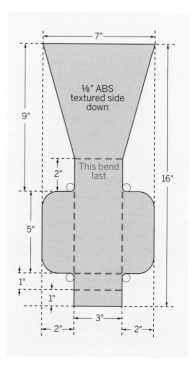

7"

⅛" ABS textured side down

9"

2"

This bend last

16"

5"

1"

1"

3"

2"   2"

The blade I use is a Freud 80T, but there are others. If you use a blade that is not suitable, it will start to accumulate smears of plastic on its flat area. This is the final warning you will get. Clean that blade with a solvent such as acetone, and never use it for ABS again.

To make long, straight cuts you can also use a panel saw (big and expensive, but safe and accurate), or a handheld circular saw guided with a straightedge clamped to the sheet. This is the method I prefer. A battery-powered circular saw has enough it less likely to melt the plastic.

For smaller cuts, a band saw is trouble-free. Since ABS is soft, you can also use hand tools with very little effort, especially a Japanese-style, pull-to-cut saw, which makes exceptionally clean cuts. When pulling it, be careful that it doesn't jump out of the cut and across your hand.

3. MARK AND SHAPE THE ABS

3a. Mark your cutouts. After you have your 7"×16" rectangle, clean its edges with a deburring tool, then place it textured-side down and use a fine-point water-soluble pen to draw the shape that you're going to take out of the piece. Afterward, you can wipe the lines away with a damp cloth. Don't use a permanent marker, as the solvents to clean it will dissolve the plastic.

Photography by Charles Platt; special thanks to Kelly Kingston for demonstrating

3b. Drill holes at the inside corners. ABS tends to open a fissure when you bend it at any inside corner where you don't have a smooth radius. Therefore, you need to make ½" holes at these corners, as shown on the plan (shown on page 102). I use a template (available from any stationery store) to mark the circles (as seen in Step 3a). If I don't take this step, I tend to forget to drill the holes. A regular ½" bit is too aggressive for drilling ABS; it will tend to jam itself into the plastic within one turn of the drill. Forstner bits work better.

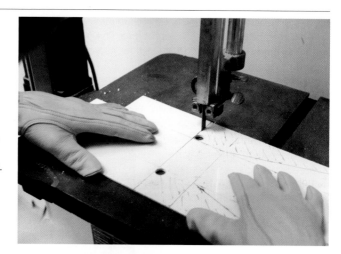

3c. Cut out the piece. After drilling the holes, cut around the edges of the shape using a band saw, hand saw, or jigsaw. The plan shows some of the outside corners rounded, in addition to the inside corners, but this is a matter of taste.

My favorite sawing tool is a DeWalt XRP jigsaw using Bosch blades, designed for hardwood or plastic. This will cut complex curves in ABS as easily as scissors cutting paper.

TIP: Saw slightly outside of each line, so that if the blade wanders you can use a coarse metal file to straighten the edge afterward.

3d. Remove cut marks and add bend marks. Remove any lines that you no longer need because they will become harder to erase after you apply heat during the bending process. Clean the plastic with a soft sponge and dishwashing liquid (never use solvents such as xylene or acetone), then make just a couple of dots to remind you where you will position each bend.

4. BEND THE ABS

This is the fun part. You need a plastic bender, which is an electric heating element mounted in a long, thin box that you place on your workbench. The bender I use is made by FTM, a company that offers all kinds of neat gadgets for working with plastic. Their cheapest bender is just over $200 with a 2' element. You can get a 4' model for about $50 more. Be careful; the bender will inflict serious burns if you happen to rest your hand on it accidentally, and since it has no warning light, you can easily forget you have left it plugged in. Gloves are definitely advisable.

Simply lay the plastic over the hot element for a brief time (25–30 seconds for ⅛" ABS, 40–45 seconds for ³⁄₁₆", and up to 1 minute for ¼"). If you overheat the plastic, you'll smell it, and when you turn it over, it will look like brown melted cheese. Naturally you should learn to intervene before the plastic reaches that point.

ABS is ready to bend when it yields to gentle pressure. Take it off the bender and bend it away from the side that you heated. If you bend it toward the hot side, the softened plastic will bunch up inside the bend, which doesn't look nice.

You can work with it for about half a minute, and when you have it the way you want it, spray water on it to make it set quickly. Alternatively, if you need more time, you can reheat it. Since the force necessary to bend the sheet increases in proportion with the length of the bend, a long bend can be difficult, so I usually apply a loose vise at intervals.

When making multiple bends in ABS, the sequence is important. If you don't think ahead, you may find that a bend you just made creates a shape that won't lie flat on the bender anymore, leaving you unable to continue. The page stand design is fairly simple, but you'll be in trouble if you don't check the plan and make the bend on the dashed blue line after all those on the dashed red lines.

When creating your own designs, it's safest to model them in paper or cardboard first.

TIP FROM THE MAKE LAB: If the $200 pricetag for a professional bender is too steep for you, don't worry, you can build one. Tap Plastics sells a bare 3' heating element, and with a little plywood, tinfoil, and fiberglass tape (also from Tap) you can make your own bender for about $65.
   Use two pieces of ¼" plywood, placed ¾" apart to make a raised channel on another piece of plywood. Then, cover the entire channel with several layers of thick tinfoil and a few layers fiberglass tape. Lay the heating element in the channel, plug it in, and start bending.
   For complete video instructions, check out **tapplastics.com**.

FINISH

NOW TRY OTHER PROJECTS 》

BENDING ABS PLASTIC TO YOUR WILL

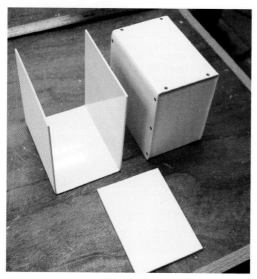

CD CADDY

One type of object that's difficult to make with ABS is a box. If you cut a cross shape (like the Red Cross logo) and then bend the 4 arms of the cross upward to make 4 sides of a box, their edges will not align accurately, and you will have no easy way to join them together — unless you try plastic welding. Some people claim they can make good plastic welds using appropriate equipment, but I've never seen a really neat weld made by hand.

Rather than try to force ABS to form conventional box shapes, it's easier to make unconventional box shapes that are appropriate for ABS. The diagram on the next page shows a plan for a very durable 2-piece traveling CD caddy. Many soft binders are sold for CDs, but I like to keep CDs in their jewel cases, and I wanted an indestructible hard-shell caddy that would let me do this.

My box-making strategy was to bend ⅛" ABS to form a 3-sided shell around thicker end pieces of 7/32" or ¼" ABS. I attached the shell by driving small screws through it, into the edges of the end pieces.

Since ABS has no grain, it tolerates this easily, so long as you drill adequate guide holes first.

Start by bending the shell to fit the end pieces. Don't be surprised if you have to deviate slightly from the lines in the plan. Mostly the plastic stretches around the outside of each bend, but a little shrinkage occurs on the inside too. You lose perhaps 1/32" on the inside of a bend in ⅛" plastic, although this may vary depending on how hot the plastic is when you bend it. Some trial and error is inevitable. If the fit is too tight or too loose, you can reheat the bend and lean on it to push it in the direction you want.

You'll need #4 stainless steel sheet metal screws, flat-headed, ⅝" long, for the next step. Assuming you have drilled holes in the shell as the plan suggests, countersink them very gently to avoid eating too deeply into the plastic, then hold each end piece in place and mark its edge by poking a pen through the holes in the shell. Remove the shell and drill guide holes in each end piece, centered within the thickness of the plastic. Because ABS does not

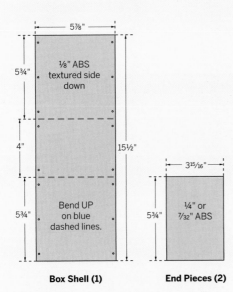

Box Shell (1)

⅛" ABS textured side down

5⅞"

5¾"

4"

15½"

5¾"

Bend UP on blue dashed lines.

End Pieces (2)

3¹⁵⁄₁₆"

5¾"

¼" or 7⁄32" ABS

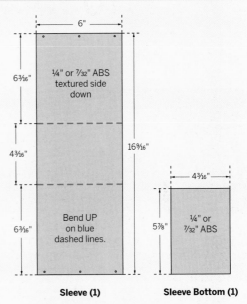

Sleeve (1)

6"

6³⁄₁₆"

¼" or 7⁄32" ABS textured side down

4³⁄₁₆"

16⁹⁄₁₆"

6³⁄₁₆"

Bend UP on blue dashed lines.

Sleeve Bottom (1)

4³⁄₁₆"

5⅞"

¼" or 7⁄32" ABS

Drill holes 3⁄32" diameter, centers 3⁄16" from each edge, offset vertically ½" from each corner or fold line.

Drill holes 3⁄32" diameter, centers 3⁄16" from each edge, offset horizontally ½" from each corner.

compress like wood, the holes must be larger than you might expect; otherwise, the plastic will swell around the screw. A ³⁄32" bit is just right for a #4 screw.

After assembling the box, the last step is to make a lid. Because I dislike hinges and catches, I chose to fabricate a sleeve that fits snugly around the box to protect its contents. The slick surface of the plastic allows the sleeve to slide on and off with a smooth, gliding action, even though it fits tightly.

Cut and bend a rectangle to form 3 sides of the sleeve, and use another thicker piece of ABS as the fourth side. Make 2 bends so that the rectangle fits around your box, then attach the fourth side of the sleeve with #4 screws using the same technique as before, and the job is done.

PEN AND PENCIL RACK

ABS makes it trivially easy to create desktop accessories. With 4 bends in a single rectangle, and some ½" holes, I created a nifty little rack so I can see immediately if anyone runs off with one of my pens! Go to makezine.com/10/abs to see a photo.

OTHER PROJECTS

Many items in the kitchen or bathroom are ideal for ABS, such as a toothbrush stand, a spice rack, or a holder for soap and shampoo to hang on the side of your bathtub. Of course you could buy these items, but by making your own you end up with something that is exactly right for your particular needs.

ABS is very flexible, but a couple of longitudinal bends will make it extremely rigid (comparable to aluminum after it has been formed into a channel or a tube). You can even make bookshelves out of ABS if you add a bend along each of the 2 long edges of a strip. Similarly, you could make a stand for your video monitor, or even a chair, if you're feeling ambitious. At my workplace, I found that after bending ABS into a channel shape, it was strong enough to support a pair of 20-lb. compressed gas cylinders.

TABLETOP BIOSPHERE

By Martin John Brown

ECOSYSTEMS ENGINEERING

The Tabletop Shrimp Support Module (TSSM) is a fun demonstration of the ecological cycles that keep us alive — and an enticement to muse on everything from godhood to space colonization.

When my 7th grade vocational aptitude test came back stamped "Forester" instead of "Astronaut," I knew the test-makers had screwed up. Sure, I liked sitting in streams, and peering down those creepy holes by the roots of old trees. But I also knew that someday the whole frickin' park would be flying through space. Hadn't anyone else seen *Battlestar Galactica*?

Now we know that space colonists are just as likely to be muddy ecologists as hotshot flyboys — the kind of people who assemble ecosystems instead of engines. Today's pack-it-in, pack-it-out life support is impractical for long, manned missions, but in the future, regenerative systems could provide years' worth of food, air, and water while processing human waste. It's recycling and reuse on a radical scale, light years beyond anything pitched by those hairy guys down at the co-op.

Here's a mini version of this dream, a sealed system that supplies a freshwater shrimp "econaut" with food, oxygen, and waste processing for a desktop journey of 3 months or more.

Set up: p.113   Make it: p.114   Use it: p.117

Martin John Brown (**martinjohnbrown.net**) is writing a book about isolation, biological and otherwise. He really likes the blog **bottleworld.net**.

Photography by Sam Murphy

TABLETOP SHRIMP SUPPORT MODULE: HOW IT WORKS

ENCLOSURE
Glass jar transmits light, but prevents materials (and inhabitants) from entering or leaving.

ECONAUT
A shrimp that consumes algae and plant matter. Like all animals, it consumes O_2 and emits CO_2.

PHOTOSYNTHESIZER
Floating plant converts light and CO_2 to O_2 and food for animals.

SCAVENGERS
Tiny amphipods, ostracods, and copepods such as *Cyclops*, eat tiny bits of plant and animal matter.

pH BUFFER
Rocks, shells, or mineral powders contain $CaCO_3$, which helps stabilize pH.

ENERGY SOURCE
Light from sun or bulb powers photosynthesis.

RECYCLERS
Microbes include photosynthetic algae and decomposing bacteria, which variously consume or liberate chemical nutrients.

CLEANING CREW
Assorted snails consume algae, clean the glass walls, and reproduce freely.

REFUGE
Kitschy aquarium ornament provides hiding place for small or stressed animals.

On Spaceship Earth, little goes in or out except light and heat, and all organisms live off each other's waste, whether it's oxygen from plants or feces from animals. Our world is bottled up.

Ecologists have often scaled down these processes, creating sealed aquariums for research. Meanwhile, space scientists have searched for organism and machine combinations that could cooperate to support humans in a space colony.

The TSSM's basic principles come from ecologist H.T. Odum, but many details derive from the Autonomous Biological System (ABS), a sealed

aquarium invented by Jane Poynter, which has returned healthy from extended trips on the space shuttle and the Mir and ISS space stations.

The Cast of Waterworld

In our TSSM, the "econaut" we imagine ourselves in the place of is a shrimp. We encourage photosynthesis and waste processing with abundant light and vascular plants, and we limit oxygen demand by constraining animal biomass and algae-fertilizing nitrate and phosphate. Protection against chemical spikes comes from pH buffers.

Illustration by Dustin Hostetler/UPSO

SET UP.

MATERIALS

[A] 1-quart glass canning jar **Don't use plastic; it may bleed air.**

[B] Clear bottles or plastic containers **for sampling and a "holding tank"**

[C] Tap water

[D] Small river rocks **just enough to cover the jar bottom. Rocks piled too thick let muck and algae build up where snails and shrimp cannot eat them.**

FROM AN AQUARIUM STORE

[E] Tap-water dechlorinator

[F] Aquarium ornament(s) or other glass or ceramic obstacle(s) **Seashells also are nice, and supply extra calcium carbonate.**

[G] Fine fishnet or kitchen strainer

[H] Freshwater minerals such as "Kent Freshwater" or "cichlid salts" **These are essential trace nutrients.**

[I] Amano shrimp (1) (*Caridina multidentata*) an algae-eater with a reputation for tolerating high pH

[J] Snails (4) of assorted species **smaller than 1cm each.**

[K] 8 stem inches of hornwort (*Ceratophyllum demersum*)

[L] 2"×2" piece of duckweed (*Lemna*). **You can also collect this from a local pond.**

[NOT SHOWN] 1Tbsp powdered calcium carbonate **This is your primary pH buffer.**

FROM A LOCAL POND

Assorted amphipods (2–8) **These are tiny crustaceans; try to collect 8, but you can use fewer.**

1 or 2Tbsp pond sludge **hopefully containing copepods and ostracods (even tinier crustaceans), bacteria, microalgae, etc.**

NOTE: Aquarium fish, shrimps, and snails may be invasive and destructive if released into the environment, so boil or freeze them after the experiment. Or keep them living in an aquarium environment.

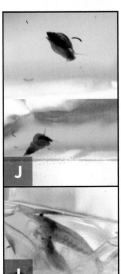

MAKE IT.

CREATE YOUR BIOSPHERE

START ››   Time: **A Day**   Complexity: **Easy**

1. GATHER THE AQUARIUM SUPPLIES

1a. Visit an aquarium store for the materials listed on the previous page. While you're there, ask them how to dechlorinate local tap water for aquarium use.

NOTE: The store staff might not believe that your Tabletop Shrimp Support Module will work. Make nice anyway.

1b. At home, dump your shrimp, snails, hornwort, duckweed, and the water they came in into an open "holding tank." I use a plastic Tupperware or yogurt container. Add some dechlorinated tap water to keep everything comfortable (alive).

2. COLLECT THE POND LIFE

Go to a local pond. Spring and summer are best. Bring a net or bottle (or other container), and visit during late afternoon. That's when the pH is higher, like that of your TSSM.

2a. Find a good, shallow area of the pond to collect your goodies. If you see duckweed, water lilies, or other vascular plants, try near there. I've done well in areas with a mixture of substrates, like sand, rock, and decaying wood.

Aquarium store and pond photos by Martin John Brown

2b. Drag your bottle or net through mud, rocks, and half-submerged plants. Examine your take for shrimp-like creatures 1mm–10mm long. These are probably amphipods; collect up to 8 of these if you can. You need to look aggressively, getting into the muck and shaking bits of plant away. Then collect 1 or 2Tbsp of pond sludge from the pond bottom, which should contain some nearly microscopic copepods and ostracods. Back home, dump your pond samples and sludge into the holding tank.

3. BOTTLE IT UP

3a. In a new container, whip up a gallon of NPFW (nitrate-poor fresh water). This is tap water, dechlorinated and supplemented with your freshwater mineral mix (follow package directions).

3b. Thoroughly rinse your "fixtures" — quart canning jar, ornaments, rocks, etc. — with NPFW.

NOTE: Waters from the aquarium store and pond are probably loaded with algae and algae-supporting nitrates, which will lead to algae takeover. Diluting with NPFW helps prevent this.

3c. Fill your jar halfway with NPFW, and transfer all the ingredients to the jar, except for calcium carbonate powder, if used: shrimp, snails, horn-wort, duckweed, amphipods, sludge, ornaments, rocks, seashells. Use the quantities listed. Do not put in extra animals or sludge, or otherwise mimic a traditional aquarium. What makes this system work is its sparseness.

3d. Fill the remaining volume of the canning jar with NPFW, leaving 1" or 2" of airspace at the top. If you have cal-cium carbonate, add it last, and note that it will cloud the water for hours.

3e. Say a little prayer as you tighten the cap on the jar.

3f. Your biosphere is complete! Place it in a spot with a fairly con-sistent temperature (70–80°F) and 12–16 daily hours of moderate light. Standard room lighting is too dim, and direct sun is too much. A bright north window or a 50W bulb a few feet away are both good, but watch the temperature.

FINISH X

NOW GO USE IT »

ENJOY YOUR BIOSPHERE

Maintaining the TSSM is a joy. There's no feeding or fiddling with parameters. Just observe and philosophize. Get enchanted with your econaut shrimp, casting its antennae in slow looping rhythms. Watch the snails cruise the glass like silent Sumo wrestlers on night patrol. Zoom in on the tiny creatures oozing out of the muck. They are the bottom of the food chain, the disassemblers of the dead.

There's never been another world like this one. In a way, you're God! Which might bring on some curious emotions if something goes awry. Multispecies assemblages like the TSSM are never 100% reliable. Your econaut might die mysteriously. Or you might observe signs of stress: shrimp that molt and then shrink instead of grow, or carnivory among normally vegetarian shrimp or snails. Hard questions arise. Was it right to start this world? Will you intervene, or abandon your creations to a sealed fate?

Life inside such tight ecological loops is rarely a cakewalk, and this begs some questions. Does closed-system sustainability simply emerge as you scale things up? Or is there something about the Earth and its milieu of flux on flux that we've failed to understand so far? Might our increasingly crowded planet, with a rising rate of extinctions, start resembling a laboratory microcosm? And for those with sci-fi dreams, could living on Mars be little more than desperate farming?

But if ecosystems engineering makes progress, we have hope. Mark Kliss, chief of the Bioengineering Branch at NASA's Ames Research Center, envisions extraterrestrial life support systems that provide a high quality of life, with a big contribution from automation. Machines and software could monitor conditions and energy inputs, nudging ecological feedback loops away from mutual parasitism and into productive symbiosis.

It's a vision our environmental movement might consider. The thing that finally allows people to live in balance with nature might be technology, the force that once seemed most opposed to it.

Amano shrimp chills upside down.

Snail grazes on algae.

Snail-on-hornwort action.

"Econaut" Jane Poynter working within Biosphere 2.

Beyond Spaceship Earth

At least 5kg of food, water, and oxygen must be lifted into space for every person-day spent on the International Space Station, relates NASA's Mark Kliss. For human habitation on the Moon, Mars, or elsewhere — stays of hundreds or thousands of days — that adds up to an unworkable ball and chain.

That's why Kliss and others are trying to replicate the closed-system sustainability of Spaceship Earth. Academics have long built "closed ecosystem" models for streams or lakes to investigate subjects like carbon cycling and population dynamics. For potential space travel, the conditions are far more constrained. Species may be mixed in ways never seen in nature, but must include the target species, humans.

American and Russian space scientists have been working on the problem since the 1960s. Early Russian tests were brutally simple: one guy climbed into a cask with little more than a light and a bucket of photosynthetic algae, to stumble out 24 hours later, alive and stinking. Progress has been slow, and no bioregenerative systems have yet been used in space for human life support.

Research has followed two paths. Space agencies have focused on highly engineered systems that include just a few well-understood species and fully account for their chemical products and needs.

Projects like Biosphere 2 (and the TSSM project here), however, take a more top-down approach. Thousands of species were imported to Biosphere 2's fantastic glass structure in the Arizona desert, and assembled into new forests, farms, and "oceans." By the time eight jump-suited "econauts" were sealed in, in 1991, it was a publicity juggernaut.

Over the next two years — the duration of a Mars expedition — the econauts met the recycling challenge, surviving very largely on regenerated air, food, and water. But their elaborate menagerie suffered a hard shakeout. Oxygen declined to dangerously low levels, and food became scarce. Extinctions were rampant and, critically, included all the pollinating species.

Life in Biosphere 2, that questing ecological utopia, wasn't sustainable. When ecosystems are sealed off, it's *Escape from New York*. Systems must balance locally, and an ecological shakeout ensues. The community that emerges may be strange and new, or as dismal as pond scum. Even with our TSSM, you can follow the same recipe to bottle up more than one tabletop biosphere, and things will evolve in different directions.

As Kliss philosophizes, closed ecosystems tread a fine line between symbiosis and mutual parasitism. Will the inhabitants help each other survive, or eat each other alive?

➕ Resources at makezine.com/10/biosphere

Vibrobot

By Mark Frauenfelder

Make a twitchy, bug-like robot with a toy motor and a mint tin.

You will need: Metal candy mint tin, wire coat hanger, 1.5V motor from a battery-powered toy, small metal washers (4), small bolts and nuts (2), about 1' of insulated wire, paper clip, ¼M flat plastic faucet washers ³⁷⁄₆₄" OD (3), AA battery, hot glue gun, hot glue, cable tie

When my 3-year-old daughter dropped the $1 battery-powered fan I bought her, the plastic case cracked, ruining it. I promised her I'd make something even better using the fan's motor. I'm a fan of Chico Bicalho's wonderful windup toys, so I made a robot inspired by his designs. I call mine the Vibrobot, and you can make one in a couple of hours or less.

1. Prepare the candy tin.

Sand the paint off the tin, if you wish. Punch 2 holes through the bottom of the tin on either end, using a hammer and a Phillips screwdriver. You'll use these holes to attach the legs. Punch a hole through the lid near one end. This hole is for routing the wires.

2. Make the legs.

Snip off 2 long pieces of wire from a coat hanger and bend each into a V-shape. Bend the tip of the V into a right angle, and then bend a little "foot" at each end (Figure A). Attach the legs to the holes in the tin using bolts, nuts, and metal washers (Figure B). Add a dollop of hot glue to each foot to give them rubber tips.

3. Install the motor.

Push a paper clip through one of the plastic flat washers, and attach the washer to the spindle of the motor. Solder 2 wires to the 1.5V battery, insert the battery in the candy tin, and thread both wires through the hole in the lid. Solder one wire to a lead on the motor, and solder a third loose wire to the other motor lead. Put 2 plastic flat washers between the motor and the candy tin, and secure the motor to the tin using a cable tie.

To operate the Vibrobot, twist the loose battery wire and the loose motor wire together (you can also solder an alligator clip to one of the wires for a switch). Experiment with the critter by gently bending the paper clip and legs into different shapes and observing the effects. Watch a video at makezine.com/10/123_vibrobot.

Mark Frauenfelder is editor-in-chief of MAKE.

Photography by Carla Sinclair

LADY BENDS THE TUBES

Shawna Peterson was in the middle of completing a degree in psychology when she landed a part-time job "babysitting the store" at a neon shop to help pay her way through school. After two years, she started an old-fashioned apprenticeship with the tube bender after work, mastering each step before she was allowed to proceed to the next. "Even after five years, you're still not a journeyman," she says matter-of-factly. Twenty years later, she's a master glass bender who divides her time between creating commercial neon for businesses, teaching neon bending at her workshop in Emeryville, Calif., and making neon artwork, whether commissioned or her own. Does her background studying cognition help? "Bending a neon pattern is like working a mental puzzle, every time," she says. "You need to plan it out, carefully and creatively."

Peterson's workshop is packed with her own work and the tools, both high-tech and humble, of her trade. Her crossfires for bending the glass tubing are pretty impressive, but she admits that her favorite tool is a 15-year-old charred wood block that she uses to cool freshly bent glass. While glass-bending technology hasn't changed much since the early days, she's the possessor of a modern, greaseless O-ring vacuum manifold for emptying oxygen and other impurities from the bent tubes and then pumping in neon and argon. Once the tubes are sealed and wired to the transformer, it's time to let there be light. After all, "There's nothing more satisfying than building something from scratch and then lighting it up."

—Arwen O'Reilly

Photograph by Robyn Twomey
+ More info: petersonneon.com. Full interview and more images at makezine.com/10/workshop.

1. Neon scrap rack for future projects. 2. Models for sculptures. 3. Antique glass "bubble" clock ("Monroe Shock Absorbers"). 4. Neon electrode box. 5. Hand torch for splicing neon tubing. 6. *Joker*, neon rendering of an antique playing card. 7. Crossfires and ribbon burner for bending neon tubing. 8. *Pimpin' Yellowjacket*, Berkeley High School mascot created for the Class of 1985.

NEON MANUFACTURING

MAKE's favorite puzzles. (When you're ready to check your answers, visit makezine.com/10/aha.)

Switch or Miss

Late one night, an evil witch rounds up 23 maidens from the surrounding village and makes them prisoners in her dungeon. The next morning she tells the maidens she will free them if, and only if, they can pass the test she has devised.

A room in the tower of the castle has two switches (we'll call them switches A and B). The witch says, "Whenever I feel like it, I will lead any one of you to the top of the tower and make you toggle one of the switches (either move it from the up position to the down position, or from the down position to the up position). You may choose which switch to toggle; however you must toggle one and only one switch, and I'm not telling you which position they are in to start with. Afterward, you will be led back to the dungeon, and I will continue selecting maidens and forcing them to toggle a switch. My selection of maidens each time will be random, meaning I may even take the same person up the tower multiple times."

"When you think that all of you have visited the tower at least once, any one of you may state that this has happened. If you are correct, you will all be freed. If you are wrong, you will be locked in the dungeon forever."

"You will have one chance now to confer with each other and devise a plan. After that, you will be kept in separate cells and have no chance at further communication."

The maidens come up with a plan using only the switches to communicate. Using this plan they will be able to say with 100% certainty that each maiden has visited the switch room. The maidens will not be able to speak to each other or communicate in any other way besides looking at the position of the switches in the tower. What is their plan?

Seat Shuffle

A line of 100 passengers is waiting to board a plane. They each hold a ticket for one of the 100 seats on that flight. (For convenience, assume that the nth passenger in line has a ticket for seat number n. For example, the first person has a ticket for seat #1, etc.)

Unfortunately, the first person in line is crazy, and will ignore the seat number on their ticket, picking any random seat out of all 100 seats to occupy. All of the other passengers are quite normal, and will go to the proper seat unless it is already occupied. If it is occupied, they will then find a free seat to sit in at random.

What are the chances that the last (100th) person to board the plane will sit in the proper seat (#100)?

Michael Pryor is the co-founder and president of Fog Creek Software. He runs a technical interview site at techinterview.org.

Illustrations by Roy Doty

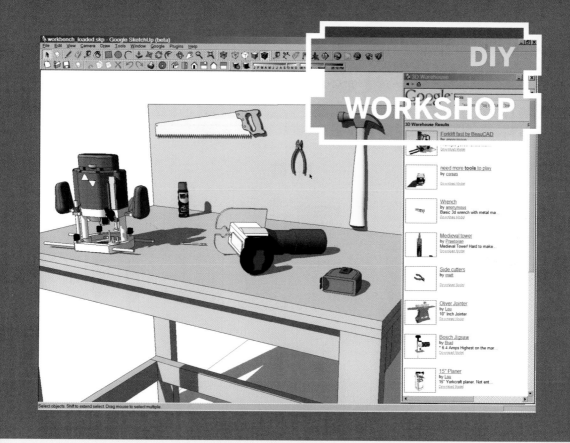

SKETCHUP WORKBENCH

Design your own work area with Google's free drawing application. By John Edgar Park

Google SketchUp is my favorite design tool, and if all goes according to plan, it'll soon be yours, too. Even though I use higher-end 3D software all day at work, SketchUp still blows me away; it enables fast, fun, and accurate 3D sketching unlike any other program (it's free too!).

Makers will find SketchUp useful for all sorts of things, from furniture design to workshop layout, from project enclosures to robotic exoskeletons. It's good for this kind of stuff because you can rough out your designs quickly, using real-world dimensions.

I decided to use SketchUp to design a much-needed workbench. The first phase was to create the conceptual model, which is a rough 3D sketch of the form. The second phase was design engineering, where I figured out the real-world materials list and construction plan for the project.

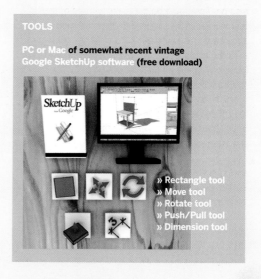

TOOLS

PC or Mac of somewhat recent vintage
Google SketchUp software (free download)

» Rectangle tool
» Move tool
» Rotate tool
» Push/Pull tool
» Dimension tool

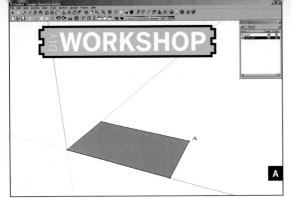

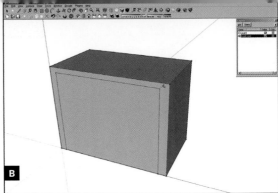

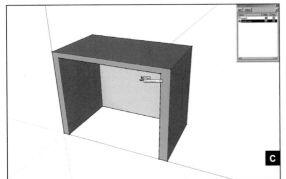

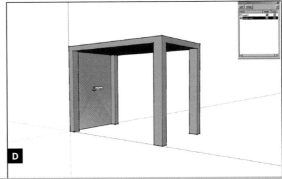

Fig. A: Draw a rectangle.

Fig. B: Mark the side for the cutout.

Fig. C: Carve out the underside.

Fig. D: Make the leg cutouts.

Build a Workbench in SketchUp

Phase I. Conceptual Design

1. Get SketchUp running.

1a. Download and install SketchUp from sketchup.google.com. It's available for OS X and Windows XP (please join me in begging for a Linux version).

1b. Launch SketchUp and do the introductory tutorial listed under Help→Self-Paced Tutorials →Intro to get a feel for viewport navigation and the basic drawing tools.

2. Prepare your project.

2a. Create a new project by clicking File→New. Set the units to fractional inches by going to Window→ Model Info, choosing the Units category on the left, and then picking Fractional from the Format list. This means that measurements in this project will be listed in inches only, instead of feet and inches. Also, go to Window→Styles, click the Edit tab, and turn on Endpoints. This makes vertices easier to see.

2b. Use the Select tool (found under Tools→Select) to click on the 2D man living in your scene. His name

is Bryce. Click Edit→Hide and wave goodbye to Bryce.

2c. I like to organize the models within each project on their own layers to control visibility and interaction between parts. Add a new layer for the conceptual phase by choosing Window→Layers and then clicking the Add Layer button in the Layers window. Name the layer Rough Layer and make it active by clicking the Active radio button. The active layer is where all new objects will go.

3. Rough out the form.

3a. It all begins with a rectangle. Choose Draw→ Rectangle. Now, click the left mouse button on the origin (the center of the scene where all axes cross) and drag toward an opposite corner, paying attention to the measurements in the lower right corner of the interface. Release the mouse button to finish. Immediately after you draw a shape, you can type in dimensions to set an exact size; type 48", 28" and press Enter on the keyboard. (No need to click anywhere, just start typing.) A shaded rectangle appears (Figure A).

3b. Extrude the tabletop upward to give the model height. Choose Tools→Push/Pull. This tool is fun to use; put the cursor over the tabletop face, then click

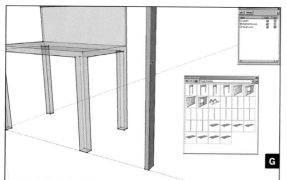

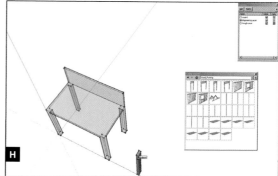

Fig. E: Measure out a line for the pegboard.
Fig. F: Paint the rough model translucent.

Fig. G: Measure for the leg cut.
Fig. H: Group the legs.

and drag upward. Release the mouse button, and key in an exact height of 36".

3c. Time to cut out the underside. Create a measurement guide 3" from the bottom left corner with the Tape Measure tool by clicking once on the corner point and a second time anywhere along the bottom edge. Type 3" and press Enter to set the exact measurement. Use the Rectangle tool to draw on the front face of the model. Start the rectangle at the measurement guide you just made. End the rectangle at around 42", 34" — again, you can type these dimensions to be precise. Although this is a rough model, some of the following steps work best if the rectangles you draw are of a consistent height (Figure B).

3d. Use the Push/Pull tool to push this new face all the way to the back of the model. You'll see an inference pop-up declare "On Face" when your cursor is aligned with the back face. Release the mouse button and you'll have carved a large chunk out of the model (Figure C).

3e. Repeat this procedure twice more on the inner sides of the workbench to leave the tabletop standing on 4 legs. Start each rectangle at the bottom edge, 2½" from the side, measuring this off with the Tape Measure tool first. The dimensions should be 23", 34" (Figure D).

4. Add details.

4a. Next, add a pegboard for tool storage. Choose the Line tool (the pencil) from the Draw→Line menu item. Click a point on the left edge of the tabletop near the back edge of the table. Begin moving the cursor to the right side to draw your line — a red inference line appears when your line is parallel to the x-axis. Continue until you reach the right edge and a message pops up to let you know you've intersected the edge. Click to lay down the second point, which will complete your line (Figure E).

4b. Using your Push/Pull tool, pull up the small face at the rear of the tabletop to an appropriate height, around 16".

4c. In the next phase, you'll use this rough model as a template for your design engineering model. To make that easier, paint a semitransparent material on the rough model. Go to Tools→Paint Bucket, and choose Blue Glass from the Transparent palette. Shift-click your model to paint it (Figure F).

4d. Save your scene by clicking File→Save As and type in the filename *workbench.skp*. Click the Save button.

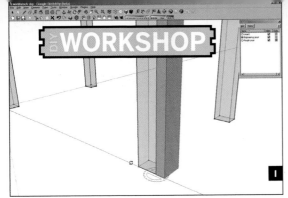

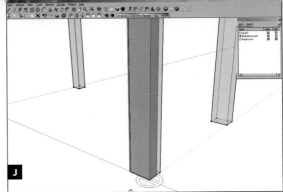

I | J

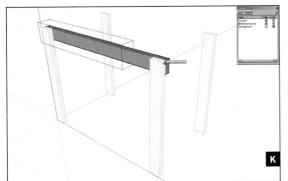

K | L

Fig. I: Prepare to rotate the leg.
Fig. J: Rotate the leg 90°.

Fig. K: Create an upper support stud.
Fig. L: Duplicate the upper parts for the lower frame.

Phase II. Design Engineering

5. Choose components.
5a. Create a new layer (Window→Layers). Name it Construction Layer, and make the layer active.
5b. Open the Component window by clicking Window →Components, and then choose Construction from the drop-down menu. Here you'll find the 2×4 we'll need. Click-drag the 12'-long 2×4 stud from the Component window to your scene.

6. Cut lumber.
6a. Right-click on the stud and choose Explode from the menu. Sorry, nothing dramatic happens, but this does let you edit the stud.
6b. You'll measure the cut with the Tape Measure tool. Create a measurement guide by clicking on a corner point at the bottom of the stud and then click again partway up the same edge. Type in the height of your leg cut, 34", and press Enter (Figure G).
6c. Use the Push/Pull tool to drag the top face of the stud down until it snaps to the Guide Point you measured, thus cutting the leg down to 34".
6d. With the Select tool, triple-click the stud to select all connected faces, and then group them by clicking Edit→Make Group (Figure H).

7. Make copies.
7a. Select the leg, then choose Tools→Rotate and rotate the leg 90° on the z-axis (the protractor should be blue). Then, click Tools→Move and move the leg into position over one of the legs of the rough model. Do so by clicking once on one of its bottom points, and then a second time on the equivalent point on the rough model (Figures I and J).
7b. With the leg still selected, duplicate it by clicking on Edit→Copy, and then Edit→Paste. Position the new leg and then repeat for the remaining legs.
7c. Clean up the screen by turning off the Rough Layer visibility in the Layers window.

8. Support frame.
8a. Paste, move, rotate, and resize (phew!) one of the legs to create the front upper support. In order to edit the length of the stud, double-click the group with the Select tool, then use the Push/Pull tool on one end face. Repeat this step 3 times to complete the frame (Figure K).
8b. Duplicate the frame parts downward to support the lower legs, about 6" up from the bottom. Do this by shift-selecting the parts and using Edit→Copy, then Edit→Paste (Figure L).

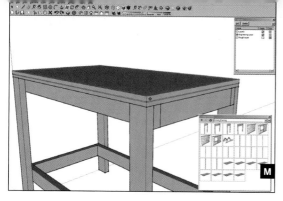

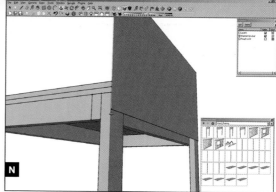

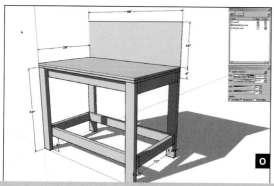

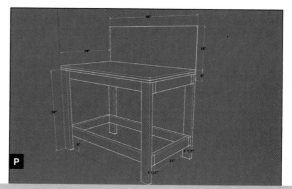

Fig. M: Cut two plywood sheets for the top.
Fig. N: Resize the pegboard.

Fig. O: View the dimensions and shadows.
Fig. P: Impress your friends with the blueprint render style.

9. Top it off.

9a. Grab a standard sheet of ¾" plywood for the top of the workbench from 3D Warehouse, the user-supported model repository. Click File→3D Warehouse→Get Models. In the search field, type "¾ thick plywood". When you find it, click Download Model, then click Yes to load it into your scene. Snap it to the top of the bench, then cut it down to size the same way you did the 2×4s. For a sturdy work surface, lay a second piece of plywood on top of the first (Figure M).
9b. You can now add the pegboard to the back. Use the 3D Warehouse to import and place a sheet of ⅛" plywood. Then, use the Push/Pull tool to extrude the pegboard piece 16" above and 4" below the bench top (Figure N). Save your model by clicking File→Save.

10. Note dimensions.

10a. To add dimension annotations to the model, choose Tools→Dimensions, then click on 2 points you need dimensions for, dragging outward to place the text. Repeat this for any unique cuts or measurements.
10b. Choose a flattering camera view and then click File→Print. Now that you're done, show off your new workbench in its best light; turn on shadow rendering by clicking View→Shadows. You can adjust lighting in Window→Shadow Settings (Figure O).

By playing with the scene's color setting found in Window→Styles, you can create the clean look of a blueprint, the loose lines of a charcoal rendering, the paranoia of a watermarked painting, and more (Figure P). Snazzy!

RESOURCES:
3D Construction Modeling by Dennis Fukai from insitebuilders.com
SketchUp Level 1 training DVD from go-2-school.com
Official SketchUp forum: groups.google.com/group/sketchup

Live by the golden rule; share your models via the 3D Warehouse. Be sure to tag them with the word *MakeMagazine*.

Looking for more models? Search the 3D Warehouse, the user repository of SketchUp models. Go there by clicking File→3D Warehouse→Get Models.

John Edgar Park rigs CG characters at Walt Disney Animation Studios. He is the author of *Understanding 3D Animation Using Maya*. Read about his house addition at parkhaus.blogspot.com.

TIGHT-FIT WORKBENCH

Make an inexpensive workspace for crowded quarters. By Todd Lappin

It's hard to be a maker if you don't have a good place to do your making. Yet two things often stand in the way of building out a basic home workbench: high cost and limited space.

Industrial-grade fixtures and spiffy garage storage systems cost a pretty penny. Likewise, domestic real estate is a scarce commodity — garages must still be used for parking cars, basements for storing stuff, and utility rooms must shelter washing machines and assorted whatnot.

I faced those constraints and a little more when I set out to build a simple workbench in my narrow garage. To avoid getting in the way of my car, the bench had to be shallow — no more than 2' deep. I needed lots of storage for tools, small parts, and bulky boxes of big stuff.

And just to make things more challenging, I also had to build the bench around several pre-existing drain and sewer pipes that intruded upon my already-limited workspace. Here's how I built the simple bench setup shown above.

Lighting

Bright, shadow-free light is essential when doing precision work or manipulating small parts. This was one area where I lucked out. We'd recently renovated our house, so my garage started out with brand-new fluorescent light fixtures running along the ceiling. Otherwise, a plug-in overhead fluorescent fixture would have been an inexpensive way to go. I also keep a simple $5 clip lamp on hand for task lighting.

Photography by Todd Lappin

The Workbench

Given my spatial constraints, I was tempted to build my own workbench from scratch, using 2×4's and plywood. Ultimately, however, I decided it was easier (and probably cheaper) to look for something off-the-shelf. The workbenches sold at many of the big chain hardware stores are overpriced and under-built, but Global Industrial (globalindustrial.com) offers several industrial-grade benches for $150 or less. Trouble is, they're also big, typically 60"×30". I didn't have that much room, and because of our intruding drain pipes, I also had to find something that didn't need to sit flush against the back wall.

I found the ideal solution at Ikea, much to my own surprise. Ikea's "Antonius" line is a cantilevered storage system built around upright metal rails that screw into the wall. A compact workbench configuration is offered, with a laminated particleboard top that's just 24" deep and 47" wide. It's sturdy, versatile, and very cheap — less than $50 for all the required parts.

Tool Storage

Ikea offers a pegboard option for the Antonius storage system, but it uses a square hole pattern that's incompatible with standard pegboard fittings. I avoided that problem by simply screwing a half-sheet of standard round-hole pegboard to the back of the workbench.

To store the rest of our tools, my wife donated the red Sears Craftsman tool chest that she'd previously used in her home office (I knew I'd married well). These are also surprisingly affordable, and basic models can be had for around $175.

Parts Storage

Ah, the little stuff: nuts, bolts, screws, nails, tapes, glues, wall anchors, wire, and whatnot. These things should be readily accessible, but storing parts in coffee cans and plastic deli containers quickly grows cumbersome.

A "pick rack" of removable plastic bins — the kind used in factories and warehouses — is a simple and affordable way to get the job done. Global Industrial sells bin unit sets that come with 32 small bins and a 36"×19" wall-mounting panel, all for around $50. Bigger sizes, with many more bins, are also available.

A "pick rack" of removable plastic bins is a simple way to organize the little stuff.

Bulk Storage

A good shelving system is the best way to make efficient use of limited floor space in garages and other mixed-use areas. Again, the temptation here is to simply build a vertical shelving unit from scratch, but in the interests of future-friendly expansion and flexibility, I'm going prefab.

The steel Gorilla Rack shelves sold at Costco or Home Depot are sturdy and cheap, but they're usually sold in just one size and configuration (which may or may not suit your needs). Global Industrial offers a variety of commercial-grade shelving systems in a very wide range of heights, widths, depths, and shelving configurations, at very reasonable prices — all the better to make the most of every square inch of precious space.

Options and Accessories

With my major infrastructure in place, I added a few more bolt-on components to complete the setup. I screwed a 2' power strip into the top-rear edge of the Ikea workbench, to provide plenty of electrical outlets for rechargeable tools and soldering irons. Magnetic strips designed for kitchen knife storage also work well to organize frequently used hand tools. Ikea sells these on the cheap, so I bought one and mounted it to the pegboard. Now I'm well-lit, neatly organized, and fully powered up. Time to get to work!

Todd Lappin (telstar@well.com) moonlights as fleet operations officer for Telstar Logistics, a leading provider of integrated services.

COOL PHOTO WEBSITES

 ## Use these online services to enhance your digital photographs. By Mark Frauenfelder

Thanks to the rise of inexpensive digital cameras and photo sharing sites like Flickr, photography is more popular than ever. I've come across a number of useful and free web-based services that make it easier to save, share, organize, and edit your digital photographs.

How can I download many Flickr photos at once?

You can batch-download multiple photos from a Flickr set with a free utility.

It's tedious work downloading a bunch of full-resolution photos from a Flickr account to your computer, requiring a lot of back-and-forth mouse clicking. Windows users have it much easier: they can grab a copy of FlickrDown (greggman.com/pages/flickrdown.htm), a nifty utility that makes

it easy to download dozens or even hundreds of photos from Flickr in one fell swoop.

After launching FlickrDown, enter the Flickr username you're interested in. After the thumbnails load, you can check the ones you want, or select "All photos" (if the user has lots of photos this could take a lot of time and consume quite a bit of hard disk space, so be careful). Then select a directory to store the files in, and click Download.

How can I immediately share photos taken with my cellphone?

Use a Blogger or Flickr account to email camera phone images to your blog or online photo gallery.

Now that the resolution of cellphone cameras is approaching that of full-scale digital cameras, some people are using them as their only point-and-shoot

Photography by Mark Frauenfelder

option. One nice thing about cellphone cameras is that you can send the photos to your blog or Flickr gallery directly. (Unless you have a plan that allows unlimited data transfer, you should check to see how expensive it is to email photos.)

To upload photos by email to a Flickr account, visit flickr.com/account/uploadbyemail, where you'll find a special email address you can use. Use the subject line of the email to give the photo a title, and write a description of the photo in the body of the email.

How can I arrange my photos visually by the location where they were taken?

Platial combines blogging, tagging, and online maps.

A website called Platial (platial.com) lets you stick virtual pushpins into a satellite photo map, and then write about the spot in its corresponding "Place-Blog." The ultimate goal is to give every square inch of the planet its very own PlaceBlog. What kind of places get pinned and blogged on Platial? Restaurants, for one. Many of the restaurants in major cities have multiple reviews, as well as stories about what happened to the bloggers who ate there. In addition to leaving written descriptions of a place, you can add photos from your computer or the web, and even upload videos. You can also create your own maps — of your favorite dance clubs, for instance — and publish them to your personal website using Platial's MapKit functionality.

How can I edit and retouch my photos online?

Use Snapshot to make quick adjustments to snapshots and other digital images.

Photoshop is overkill for people who just want to enhance, resize, and adjust their digital photos. Lately, I've been using a nice little web-based photo editor called Snapshot (snapshot.com) to upload digital photos and tweak them to my heart's content.

One of the nice features of Snapshot is its ability to import images from Flickr (or any other website) using a bookmarklet. Once you drag the bookmarklet into your browser's bookmark bar, you can go to any web page, click the bookmarklet, then choose the photo you want to edit in Snapshot.

Here's what you can do in Snapshot:
- **Resize.** Click one of the red squares and drag your mouse. Select a corner square to preserve

A

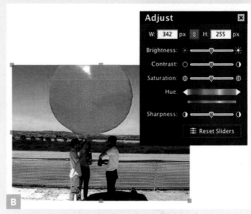

B

the size ratio of the original image (Figure A).
- **Crop.** Drag the crop window around the image, and resize it by clicking and dragging a red square. Once you're happy with the crop, press enter or double-click the image (Figure B).
- **Enhance.** Bring out the color of an image with a single click. If you aren't pleased with the result, click "Undo."
- **Adjust.** Change the size, brightness, contrast, saturation, hue, and sharpness of the image.
- **Rotate.** Click to rotate the image 90 degrees clockwise.
- **Save.** You can save the photo in a variety of formats, or upload it to your Flickr account with one click.

Excerpted from *Rule the Web* (St. Martin's Press, 2007) by Mark Frauenfelder. Visit ruletheweb.net for more tips and tricks.

Mark Frauenfelder is editor-in-chief of MAKE.

THE SWEET SOUND OF PARTICLEBOARD

Beef up the tone of open-back amps with a little thrift shop help. By David Battino

After transforming a record player and some plumbing parts into a spinning speaker (*see MAKE, Volume 05, page 24*), George "the Fat Man" Sanger is back with a new way to enhance your guitar sound.

His Goodwill Amp Enhancer is a DIY version of the commercially available Enhancer, which beefs up the tone of open-back amps by redirecting the "lost" sound to the front.

The nicely finished commercial versions start at $150 (soundenhancer.com), but the Fat Man built his enhancer out of a $15 computer desk he scavenged from a thrift shop. "It took just an hour or two," he reports, "and adds *wonderful* tone to my amp."

How It Works

The Sound Enhancer site details the science involved, but in general, the Fat Man explains, an open-back amp is a design compromise.

"In theory, a perfect speaker box would be a speaker mounted in the middle of a wall of infinite size, because that would let the sound from the front reach your ear without having been partially cancelled by the inverse sound from the back," he says. "Mom won't let us build anything infinite anymore, not after what happened last time, so we approximate the infinite wall by putting speakers into sealed boxes, also known as *infinite-baffle enclosures*.

Photography by the Fat Man

"Unfortunately, infinite-baffle enclosures make it really hard for the speaker to move, so the sound is quieter. And of course, quietness is not very rock 'n' roll, is it? So designers make a lot of amps louder (and a little funny-sounding) by opening the backs.

"This speaker stand bends the back sound around a corner, which makes it even less like the front sound, and then sends it out the front, where its slightly altered power is added to your already Majestic Volume in a rich and tonally pleasing way."

In addition to reinforcing the sound, the Goodwill Amp Enhancer points the amp at your head, letting you hear yourself louder than, and before, your bandmates do. That helps you play better, and your bandmates don't hate you for playing too loud.

Making It

1. Ready the donor. Pull the sides off the donor, place them on the floor, and lay your amp on its side in the tipped-back position you want it to sit on the stand. The back of the stand needs to rise above the opening in the back of your amp to seal it off, but it mustn't block any essential controls. The bottom-front edge of the amp will come right to the front of the stand.

2. Make the side panels. Mark the outline with a Sharpie, and saw along the resulting L-shaped line. Now your side panels are done, and they should look something like that one particularly odd block in the game Blockhead!

3. Make the floor and back wall. Make a floor and back wall for the amp by hot-gluing the other wood between the two side panels. You may need one additional narrow piece to bridge from the top of this back wall to the spot on your amp where the open back stops and the controls begin. Don't worry if you mess up; hot glue can be broken free and redone easily.

4. Make it permanent. Once it looks right, make it permanent by sinking some screws in from the sides. Caulk up the cracks, then glue the weather-stripping to the edges that will touch your amp. "This stand will make your amp sound so much better," the Fat Man promises. "It has to be heard to be believed."

♪ Hear the Goodwill Amp Enhancer:
makezine.com/10/diymusic_amp
➕ More from the Fat Man: fatman.com

Sound comes out the back of the amp and is forced out the front of the stand. Beyond that, the shape of the barrier isn't too important. Be sure that the horizontal bit hits the amp's back above the speaker opening and below any controls.

David Battino is the co-author of *The Art of Digital Music* (artofdigitalmusic.com) and digital audio editor of O'Reilly's Digital Media site (digitalmedia.oreilly.com).

SOLAR-POWERED BIKE GPS

Green handlebar navigation from recycled parts. By Brian Nadel

I've spent much of my adult life dealing with either computers or bicycles. Writing about computer technology has put food on my family's table and a roof over our heads, while riding helps me unwind, clearing my head of the jargon that accumulates throughout the workday.

During the summer, I'll disappear for hours on long rides to nowhere and back. But I have to admit on some rides I've gotten so lost I have trouble finding my way home. Happily, I was able to build a solar-powered GPS mapping machine, mostly from old computer parts and software I had sitting around my office. I've seen motorcycle-mounted GPS navigation screens, but have never come across one on a bicycle, even though it seems like a natural mix of appropriate and functional technology.

The navigational screen I made not only shows me where I am, but also reminds me when it's time to head home, and leads me along the quickest route if I'm in a hurry. It even speaks directions to me, and will also play music. Here's how I put it together.

Cheap to Keep

Car GPS units can cost $1,000, but my bike GPS cost me next to nothing because I had most of the parts on hand already. (To my wife's chagrin, I keep bins of old, working parts for possible future DIY projects. To me, this is the essence of recycling!) If you don't have the parts, you can probably find them for about $150 total by combing through eBay and closeout retailers. It's all out there, and probably cheaper than you think.

Photography by Brian Nadel

Because I ride where there are a lot of trees and buildings, I also added an external antenna that attaches to the GPS via cable. When everything is together, turn the PDA on and choose the manufacturer of the PDA (Toshiba in my case).

Ready to try it out? The Ostia screen starts out by showing a red unhappy face at the bottom, which means the GPS receiver is off or not working. Select the menu item GPS→Enable GPS to turn it on, and go outside to a place with a clear view of the sky.

Expect the GPS receiver to take as long as 5 minutes to start up and plot your initial position; a yellow neutral mug at the bottom of the screen means the receiver is searching for satellites. After that, the device will update its position every few seconds.

From the menu, choose GPS→Sat. Info to plot the satellites you're seeing on a polar plot with you at the center. A bar graph at the bottom indicates each satellite's signal strength, and the green smiley face shows that everything is working. Three satellites will allow your position to be mapped (look for the red arrowhead on GPS→View Map), and with 4 or more, you'll also see your altitude. Move around and the arrow will follow you like a digital shadow, and even change direction as you do.

Let the Sun Shine In

If I wanted to rely just on the PDA's battery, the job would be almost done. But instead of being tied to regular recharging, I wanted to go off the grid. So, I plugged my GPS into a solar charger, and Velcro-mounted the panel onto my bike's front fender (Figures A and B).

To mount the PDA on the bike's handlebars, I used the circular mounting ring from an old bicycle reflector. The bracket was too loose, so I wrapped electrical tape around the handlebar to make it a little wider. I carefully bolted the mounting ring to the car dashboard PDA holder that came with the GPS receiver (Figure C), but there are dozens of other holders you can use.

This important connection can be a little tricky and require some improvisation; I used a lock washer and a spot of Locktite. Whatever you do, make sure the mounting is firm and won't loosen over time.

I put matching Velcro strips onto the PDA holder and the back of the PDA, and Velcroed the external antenna to the side of the front fender. Even though it was a mess of wires, it was time to test it. I rolled

Every combination of bicycle, PDA, GPS receiver, and solar panel will be different, so expect to improvise a little here and there. As with riding, getting there is half the fun. Components in hand, the whole project should take about 3 hours, not including a test ride to make sure it all works and that the screen doesn't fly off on the first New York City pothole.

PDA + GPS = Mobile Maps

Ironically, I needed a Windows XP computer to get my bike on the road. I installed the driver and the Ostia mapping software included with the GPS receiver, then used Microsoft's ActiveSync program to install the software on the PDA.

Finally, I picked and uploaded the maps I wanted. I selected the 10MB map set for Westchester County, N.Y., on my computer, and then re-synched the PDA to load the maps onto the flash card in the PDA's SD slot.

With the software chores out of the way, it was time to put the hardware together. I plugged the GPS receiver into the PDA's CF slot, making sure it seated firmly.

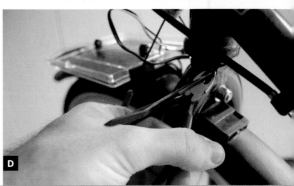

Fig. A: Plug your GPS into the solar charger.
Fig. B: Solar panels mount easily to the bike's
front fender with Velcro.

Fig. C: Bolt a car dashboard PDA holder to the handlebar,
then Velcro the PDA in there. Fig. D: Tuck the wires away
with cable ties, and you're rolling.

the bike outside, turned on the PDA, and checked its battery screen to see if the solar panel was working. Everything checked out, so I proceeded to tuck the wires away with cable ties (Figure D), leaving a little slack in every run to accommodate road bumps and turns. You can also use Velcro straps to make the equipment more easily removable.

Music and More

My first ride lasted only a couple of hours, but was eye-opening. The GPS receiver and PDA tracked me every pedal of the way, even when I went off-road. I started with a half-charged battery, but it was sunny and I returned with it almost fully charged. (When it's overcast or at dusk, the solar cell augments the battery's power.)

On the downside, the maps were sometimes barely readable while I rode over bumpy surfaces, and direct sun washed out the Toshiba e740's screen, even though it was designed for outdoor use. A greater disappointment was that, unlike newer GPS devices and programs, the Ostia software doesn't calculate speed and mileage.

But the rig is a natural conversation starter. I've been asked, "What's that?" and "Does it get cable TV?" One time I stopped to fill a tire with air and

wound up showing a car mechanic how it worked. He asked me if I used the PDA's digital music player as a two-wheeled iPod. That got me started grooving to tunes while I pedaled. I started using the device's appointment program to play an alarm when it was time to turn around and head home. Finally, I had a brainstorm and used the device's built-in wi-fi data radio to surf the web and listen to internet radio when I stopped at a local Starbucks for a break.

All told, this bike project went far beyond its original intent, opening up a new world of biking and computers. Some might call it technological overkill, but it's liberating not to worry about where I am or how to get home. Now, I only get lost when I want to.

Brian Nadel is a writer based north of New York City, and is the former editor-in-chief of *Mobile Computing & Communications* magazine. A 25-year veteran of technology journalism, he has worked for *Popular Science*, *PC Magazine*, and *Business Tokyo*.

Kipp Bradford assembles the XGameStation Pico at the Providence Geek Dinner.

BARE METAL GAME DESIGN

Introducing the XGameStation Pico.
By Brian Jepson and Kipp Bradford

Photography by Brian Jepson

André LaMothe's creations have a neo-retro bent that's hard to resist. Combining the sensibilities of game systems from 20 years ago with the DIY appeal of a microcontroller board, LaMothe's XGameStation Micro — a compact video game hardware kit — gave hobbyists the opportunity to write games that are closer to the bare metal than most programmers have been in decades.

The XGameStation Pico Edition 2.0 (makezine.com/xgamestation) takes you even closer. Modern game programming environments use collections of code libraries and high-level design tools to hide the complexity of the hardware from programmers. The Pico lets you duplicate the experience of writing games for a retro system like the Atari 2600; hardware and software fuse into a single platform, and in pushing the limits of that platform, you chal-

lenge yourself to come up with creative hacks that you would never need on today's ultra-powerful systems. Want to draw something on the screen? You'll have to understand something about video signaling first.

You shouldn't be intimidated by generating your own TV signal, because the Pico's accompanying CD-ROM includes a PDF version of LaMothe's *Design Your Own Game Console*, with more than 100 pages of new material devoted to the Pico. As you make your way through this book, you'll not only learn how to program the Pico, you'll also learn a lot about how sound and video are generated.

Unlike the Micro, you'll need to assemble the Pico Edition yourself, using the included components (building the Pico is also covered in the PDF). This can be as easy or as hard as you want,

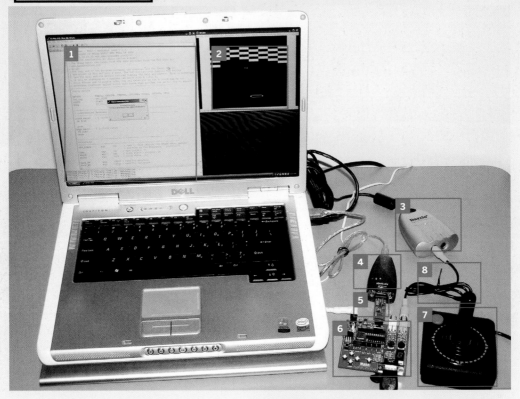

THE XGAMESTATION PICO EDITION

Shown here fully assembled and connected to a PC running the SX-Key integrated development environment (IDE).

1. The SX-Key IDE, which is included on the CD-ROM (newer versions are available for free from parallax.com)

2. DScaler, an open source video capture package (deinterlace.sourceforge.net)

3. A USB video capture device so you don't need to carry a TV set wherever you go

4. Keyspan USB/serial adapter

5. SX-Key programmer

6. XGameStation Pico Edition 2.0

7. Atari joystick

8. Composite video cable

MATERIALS

SX-Key serial programmer $50 from parallax.com, or $30 from xgamestation.com when purchased with the Pico. This lets you flash and debug SX-series microcontrollers from Parallax. (The Pico uses an SX-28.)

USB-serial adapter such as the Keyspan USA19HS ($40) if your computer doesn't have a serial port. Although the SX development tools are Windows-only, Jeff and Jason Tranter have done some work on getting the tools to run under Linux (see ca.geocities.com/jefftranter@rogers.com/xgs/index.html).

9V power supply 500mA, 2.1mm jack, positive tip. You'll need this because the onboard SX28 CPU uses a lot of power when you're flashing it with the SX-Key Serial Programmer.

because the Pico includes both a breadboard and a circuit board. So if you'd rather not pick up a soldering iron, you can put it together quickly and easily (and if you change your mind later, you can still put together the circuit board version).

However, you really have no excuse not to solder

together the circuit board version; the CD-ROM also includes a detailed video that shows how to put the Pico together. (To get your soldering skills up to speed, check out the MAKE Video Podcast soldering tutorial at makezine.com/go/solder.)

After you've put together the Pico, you can

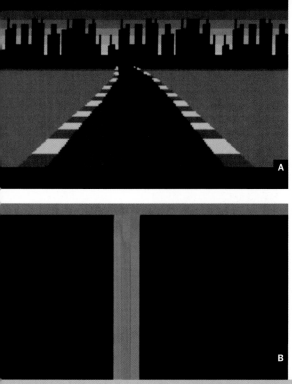

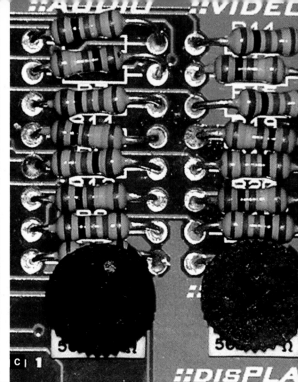

Fig. A: The Racer City Demo running on the XGame-Station Pico Edition. There's no object to it. You just drive toward a city that never gets closer — kind of like a *Twilight Zone* episode (and it's in black and white, too).

Fig. B: Generating color signals on the XGameStation Pico. Fig. C: The R2R ladder that performs the digital-to-analog magic uses only 2 resistor values.

connect a 9V battery, hook up the video cable to your TV set, and fire it up. You'll find one program on there, the Racer City Demo. This is a good showcase of the Pico's capabilities: it's a black and white program with animation and smooth scrolling. To run other programs on the Pico, you'll need to pick up a few other things, listed here in the Materials list.

The CD-ROM includes a number of example programs in the *XGSME_Sources* directory. The files with _pe_ or _pico_ in their filename are for the Pico. A number of other games are designed for the Micro Edition. To run them on the Pico, load the .src file in the SX-Key IDE, set the Pico's programming mode switch to "KEY", and select Run from the Run menu to program the Pico and execute the program.

Some of the examples override the crystal oscillator and directly control the SX-28's clock rate using the SX-Key programmer. This allows the Pico to run at a wider range of speeds. For example, this NTSC color bar demo (Figure B) requires a clock speed of 49.95MHz to properly transmit the color burst signal that embeds color information. So if you want to run one of these demos, it will have to be under the control of the SX-Key. (If you were to load the program on the Pico, switch off SX-Key mode, and restart the Pico, it would run at the wrong speed.)

The Pico uses an elegant arrangement of resistors to convert 8 of the microcontroller's digital outputs into 2 analog signals: monaural audio and NTSC or PAL video. This arrangement is known as an *R2R ladder* (Figure C). Only 2 resistor values are needed, R and 2*R (which are 180Ω and 360Ω on the Pico), and these resistors are connected like the rungs and sides of a ladder. Switching on a given digital output applies approximately 5V to the corresponding rung on the R2R ladder, with each rung adding twice the voltage of the rung below it to the analog signal. When you switch different rungs of the ladder on and off quickly, you generate the analog video and audio signals.

Once you've gotten the example programs to run, you can modify them to create programs of your own. If you start hacking your own code on the Pico, you'll find the XGameStation Forums at xgamestation .com/ubbcgi/ultimatebb.cgi very helpful. And if you write something cool, be sure to let people know about it!

Brian Jepson is a writer and hacker for MAKE. Kipp Bradford is an engineer, inventor, and volunteer at The Steel Yard. Both volunteer at AS220, a nonprofit arts center in Providence, R.I.

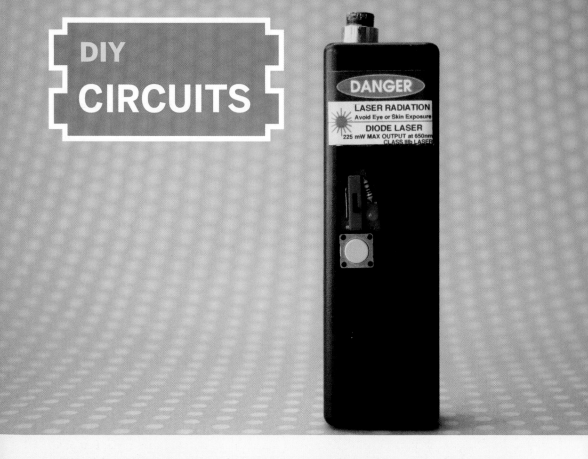

MINI HIGH-POWER LASER

 Liberate a 200mW laser from a DVD burner.
By Stephanie Maksylewich

High-speed DVD burners are mass-market commodities now. Most new computers sport DVD±RW drives, and discount shops sometimes carry upgrade burners for less than $30. They're cheap, but inside every one lies a hidden secret that many of us would have killed for 20 years ago: a high-powered, solid-state, visible red laser. This means that with a pretty small expenditure and a little hacking, you can have a portable, handheld laser that's powerful enough to ignite matches, burst balloons, and melt plastic. Here's how to do it.

Find a Donor DVD Burner

Look for a DVD burner, either single- or dual-layer, that's rated at 16x or higher. The faster the drive, the stronger its laser diode. The color will be the same regardless — DVD players and burners use a red laser at a wavelength of approximately 650nm, which is about the same as most cheap laser pointers.

Don't try this with a CD burner. These use invisible infrared light (wavelength of 784nm), which will burn your retinas out without you even seeing it. DVD burners also have CD-burning diodes inside, but we'll ignore them.

If you are looking for something unique, try a violet laser diode from an HD-DVD or Blu-ray drive. These burners are incredibly expensive right now, but they do contain a 405nm diode rated at around 50-60mW. It won't be able to pop balloons or burn things, but handheld lasers with this color light are rare at the moment. Just be advised that the 405nm diodes take a different voltage from the diode we'll be working with (5V DC rather than 3.3), so you'll need a different circuit.

MATERIALS

High-speed (16x or higher) DVD burner For this article, I used a Super Multi DVD Drive from LG, model GSA-H10N, purchased new for about $45.

Laser heat sink and collimating lens Digi-Key part #38-1000-ND ($18, digi-key.com), or else you can get AixiZ Lasers' 10mW laser module, part #AIX-650-10 (mfgcn.com, $12), and remove the diode and power supply.

Pushbutton/momentary switch

Stranded circuit wire

Small project enclosure box

AA alkaline batteries (2) in holder

OPTIONAL for better build:

Small, generic, perforated circuit board, 1N4001 diode, LED, 1kΩ resistor, SPDT (3-position) slider switch, NiMH rechargeable AA batteries (3), DC jack for external battery charger

TOOLS

Laser protective goggles for 650nm, optical density 1.5+ available from wickedlasers.com or noirlaser.com. They're expensive ($50+) but necessary; welding goggles, sunglasses, or squinting are no good. Cheap alternative: Wear a patch over the eye you want to keep.

Screwdrivers jewelry and small sizes

Soldering and desoldering equipment

Hobby knife

Dremel or other means of cutting and shaping holes in project box

Pliers

Hot glue gun and glue

OPTIONAL if you're using the AixiZ module:

A second pair of pliers, vise, hammer, small dowel, or nail set

---

⚠️ **WARNING: Never aim a laser toward a person or another living thing.** Any laser can cause harm, but this one is far more dangerous than the widely available <5mW laser pointers you're probably used to. The laser diode in a DVD burner is a Class 3b laser that produces between 150 and 200mW of laser energy. This is enough to cause instant blindness if directed into an eye.

When performing tricks, you and others watching must use proper eye protection. Reflections off shiny surfaces can be as dangerous as direct hits.

Eye damage from laser light is not always apparent right away — repeated exposures can cause a slow but permanent vision loss.

See felesmagus.com/pages/lasers-safety.html.

Disassemble the DVD Burner and Identify the Laser Diode

All the DVD burners I've worked on can be taken apart the same way. First, remove the tray cover and front bezel. Then turn the unit upside down and remove the 4 screws that hold the bottom cover in place. Unclip the circuit board and unplug any cables, then set the circuit board and bottom cover aside. You should now be able to see the optical carriage, which slides along a rail (Figure A).

Inspect the optical carriage and look for the laser diodes. They are typically two 5.6mm cylinders that are sunk into the carriage's metal chassis. Their 3 pins will face out and be connected to the carriage circuitry by small ribbon connectors.

To find the visible red diode, test both diodes in place with a pair of alkaline AA batteries in a little holder. See Figure B for the laser diode pinout, and be careful to touch the positive and negative battery leads to the corresponding pins only. All the DVD burners I've played with have used this pinout, and I hope it's an industry-wide standard, because reverse-polarizing a laser diode can fry it instantly.

Touch your battery leads to the diodes' contacts. When you've lit up the visible red one, mark it with a felt-tip so you can identify it later (Figure C).

Remove the Laser Diode

Unscrew and uncable the optical carriage. You might continue disassembling the rest of the DVD burner, which is a good source of small motors and parts for other projects. The optical carriage itself contains some goodies as well, such as beam splitters and tiny mirrors and lenses (Figure D).

Use a wick or other desoldering method to remove the ribbon cable from the laser diode's pins. Excess heat is another way to destroy a laser diode, so this is a delicate process. Use a higher soldering iron setting for short durations, quickly melting solder and then pulling away, rather than using a lower heat setting and keeping the iron in contact longer.

Methods for freeing the laser diode from its carriage will differ from burner to burner. The diode in my LG drive was held in by glue and friction, so I scraped the glue away with a knife and gently pried the diode free with a jeweler's screwdriver. With other hardware, I've had to desolder the diode. Whatever is required, be patient. Physical damage to the diode can also render it inoperative.

Once the diode is free, put it somewhere safe, and

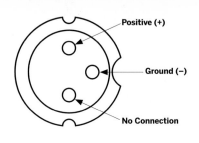

Positive (+)

Ground (−)

No Connection

Laser Diode (viewed from bottom)

Fig. A: The optical carriage slides along a rail next to the spindle, and is fed by a wide ribbon cable. Fig. B: Laser diode pinout, looking up from the pin side.

Fig. C: Find the visible red diode in the carriage by testing with a two-AA battery holder. Fig. D: Optical carriage removed from drive.

try not to touch the emitter window in front (Figure E).

Laser Diode Housing

On its own, the laser diode makes a poor laser. Its beam spreads, and it will overheat quickly. You need a collimating (focusing) lens and a heat sink. I've found two good and reasonably priced options for this: a brass housing from Digi-Key, and a laser module from AixiZ; see Materials list on previous page.

The Digi-Key housing is easier to build with but overheats more easily, while the AixiZ module has a better heat sink and costs less, but requires some surgery. The focusing action also differs between the two. The adjustable lens on the Digi-Key is tighter, and once you get it focused, it stays in position. The AixiZ lens adjusts more easily, so normal handling can spin it out of focus, but this also lets you quickly change the focus if you are performing burning/popping tricks at different ranges.

Using the Digi-Key Housing

The brass Digi-Key housing looks good (Figure F), comes with printed instructions, and is very easy to use. But it doesn't have much mass, so it can overheat, which causes it to lose focus. I've found

a duty cycle of about 1 minute of continuous use before the laser needs 1–2 minutes to cool off.

The only difficulty I had was fixing the diode in place according to the instructions. I imagine that if you already own a "diode press," as referred to, you are way ahead of me and this article. I tried to fold the mount's brass lip into place with a small screwdriver, but it wasn't working, so I just soldered the diode directly into the heat sink.

Using the AixiZ Module

This is a complete low-power 650nm laser diode module. For this project, you need to tear out and discard the 10mW diode and power regulator tucked inside. The AixiZ assembly (Figure G) is actually cheaper than the Digi-Key unit, and it has enough mass that it doesn't overheat after 3 minutes of nonstop lasering — which is the longest I've run one.

Here's the method I use to remove the original diode. First, unscrew and remove the lens and the small spring underneath, and set them somewhere safe. Unpeel the laser safety label, and use 2 vise-grip pliers to unscrew the 2 halves of the housing. You can see the seam where they join, so it's just a matter of applying enough force to break the silicone so that you can twist the halves apart (Figure H).

Photography by Stephanie Maksylewich

E F

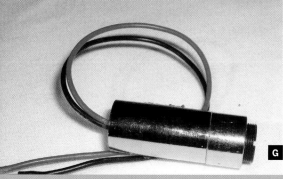

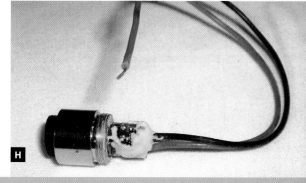

G H

Fig. E: Laser diode, removed from its carriage.
Fig. F: Digi-Key brass laser diode housing — easy to use but can overheat.

Fig. G: AixiZ 10mW laser module — inexpensive but requires some mods. Fig. H: AixiZ module cracked open, with power regulator board exposed.

Clean out the back half of the assembly with an X-Acto knife or needlenose pliers. To get the 10mW diode out of the front half, clamp it in a vise with the laser pointing up, then stick any handy plastic or wood dowel that will fit in the hole, and give it a few sharp taps with a hammer. The diode should pop free. Note that it usually dies during extraction, so you should just throw it away. Then clean out all the white silicone goop inside the front half.

Mounting the Diode into the Housing

Fit your high-power diode into the housing slowly and evenly until the back of the can is flush with the back of the heat sink. I go around the edges, pressing it from all sides with a pair of slip-joint pliers. It is very important that the diode is not crooked, and that you only press the edges of the diode can, not the center where the leads are.

Once the diode is seated, solder different-colored leads to the diode pins. Then reassemble the mount with the lens in place (Figure I). Give the laser diode another test, and focus/collimate the beam.

Circuit Details and Diagram

DVD burner laser diodes are rated for 3.3V DC and current of about 300mA. Exceeding either of these will shorten their life or cause instant death, so we need to build a power supply circuit that regulates voltage and current. I've gone two ways with this. The quick-and-dirty solution is to just use 2 AA alkalines and a switch — an arrangement I've seen in some commercial high-power red lasers. I built one laser this way, and it has worked fine. But if you use pumped-up batteries like lithiums rated at 1.7V, you'll probably toast the diode.

A nicer design is the circuit diagram in Figure J. A 3-position slider switches between Off, Continuous On, and Pushbutton-Controlled for both the laser and an LED status indicator. I have built a few lasers with this circuit, and they all still run after many months.

Power comes from 3 NiMH batteries, which can be charged in place through a jack. Rechargeable batteries typically produce about 1.2V each, so I use 3 cells with a standard diode (1N4001), forward biased, to drop the voltage a bit. I trickle-charge overnight by plugging in a wall wart adapter rated at 4.5V DC/200mA. With this arrangement, 3 NiMH AA's with 2,000mA-hours of capacity will power the laser for several weeks of running a few minutes per day.

There are lots of other ways to power the laser

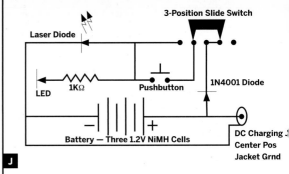

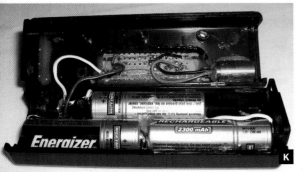

Fig. I: Laser diode wired in Digi-Key housing.
Fig. J: Schematic for laser circuit, rechargeable and switchable between continuous and pushbutton modes.

Fig. K: Laser, circuitry, and batteries hot-glued into plastic enclosure. Fig. L: The finished, pocket-sized high-power laser.

diode, but the main thing is to never exceed 3.3V DC or about 300mA. Test your circuit with a multimeter to make sure.

Assembly

The final design of the laser is limited only by your imagination. I based my pocket-sized design on the smallest off-the-shelf plastic hobby enclosure box I could find that held all the pieces.

I built my circuit on a small hobby circuit board, and then used hot-melt glue to mount my batteries, circuit board, and laser housing into my enclosure. The compact enclosure didn't have room for a AA battery holder, so I soldered and mounted the rechargeable cells inside, and exposed a DC jack that lets me recharge the batteries via an external charger. The simple circuit has no fancy controllers or drivers, so it's trivial to assemble, and these lasers can be quite compact (Figures K and L).

High-Power Demonstrations

Your laser should be able to ignite a match or burst a balloon, but there are a few things to keep in mind. Red laser light will be largely wasted on things that are colored red, orange, or some shades of yellow, and no visible laser will affect white surfaces or

highly reflective objects like metallic Mylar balloons. Black surfaces absorb the energy, so use a black marker to blacken match-heads and to make a target dot on balloons (or use black balloons).

Focus also matters; adjust it to suit the range you are testing at. A pinhole-sized dot will burst a black balloon instantly, and spreading the dot can slow the process down for dramatic effect. If you have no success, try a different brand of balloon or match.

To feel the power of the laser directly, put a bit of black electrician's tape on the back of your hand, and shine the laser on it. You should feel a sting within 1–2 seconds. This is the only time I would ever suggest directing a laser at any living thing.

Again, play safe! Even a bare diode, unfocused, can cause damage. Always direct the diode away from yourself and any reflective surfaces. If you plan to use your laser in public, check local laws first. Owning a homemade high-powered laser is not illegal in most places, but using it to cause trouble, be a nuisance, or cause blindness typically is.

As a child, Stephanie Maksylewich (felesmagus.com) took her toys apart to find out how they worked; today she does the same with all her electronic equipment.

Photograph by Brian Jepson

MAKING IT WITH THE MAKE CONTROLLER

 Our board does art, robotics, music, and more. By William Gurstelle

The MAKE Controller Kit is a powerful and easy-to-use hardware platform that can interact with the physical world. It's based on a microcontroller, which is essentially a computer-on-a-chip. Unlike general-purpose microprocessors, here the memory and device interfaces required to run a simple (or not-so-simple) application are integrated onto a single board.

The idea for the device was conceived amidst the fire and smoke that issues from the ramshackle compound where world-famous Survival Research Labs pursues its strange and violent mechanistic art. Anyone who has attended an SRL artistic review (also known as "the most dangerous shows on

Earth") never forgets the experience. The actors are machines — big, powerful, and dangerous machines.

The MAKE Controller traces its lineage to the digital controllers used to control SRL's robots and artistic weaponry. Engineers Michael Shiloh and David Williams designed much of the hardware that controlled the movement of SRL's robots and gave the art its intelligence.

We asked members of the maker community to share some of their applications. We heard from people who use the device to do an amazing number of tasks: everything from building interactive video art installations with middle-school students, to streaming selected information from the internet,

Fig. A: Because it's got built-in Ethernet and is fully programmable, the MAKE Controller is great for networked projects such as this RSS reader.

Fig. B: This balloon controller is based on the diVA project (makezine.com/go/diva), which hides the complexity of devices such as the MAKE Controller and makes it easy for people to control things through a web interface.

to controlling haunted house special effects, coffee makers, and pointing statues. Music and robotics are two areas particularly rich with interesting applications.

Sparky 2

Maker and robot expert Marque Cornblatt has been developing his interactive video chat robot, Sparky 2, over the last several years. He's incorporated the MAKE Controller board to operate the robot remotely via the internet. In robotic parlance, Sparky 2 is a "telefactor," projecting Sparky's operator's face, eyes, and voice anywhere there's an available wi-fi connection.

Inside Sparky, a wi-fi-enabled Mac mini uses the MAKE Controller to control a number of servo motors that, in turn, control the two 24-volt DC motors powering its wheels. The MAKE board's function is to respond in real time to the operator's commands. It controls the motion of the robot's wheels, but that's not all. The microcontroller watches 5 infrared sensors located around Sparky's perimeter. If the sensors spot an obstacle, the MAKE Controller enables the robot to avoid it.

Cornblatt is not done exploring the capability of the device. "The robot is under constant development," he says. "I'm working on a patch that will allow the robot to sense its remaining battery power and automatically drive itself to a charging station without requiring any human intervention in the process." The robotics designer is planning to add even more autonomous behavior in the near future.

The Bovalve

Last year, Nate Bliton, a music composition major at Michigan State University, decided to try something unique for a class project. A musician familiar with both stringed and brass instruments, Bliton developed an instrument that combined the best features of both types of instruments into a single, hybrid device. He wanted to take the control modalities of a trumpet (its valves) and combine them with the sonorous, dynamically agile capabilities of an instrument played with a bow, such as a violin. The result was a new invention, the Bovalve.

The Bovalve utilizes a multi-button joystick for the player's left hand. Three of the joystick's buttons correspond to the valves on a trumpet. But the Bovalve adds some interesting features to the standard brass player's repertoire of capabilities. Bliton uses the two additional buttons on the

Photograph of RSS reader by Brian Jepson; photograph of balloon by Shaun Mavronicolas. 2C Visual Communications

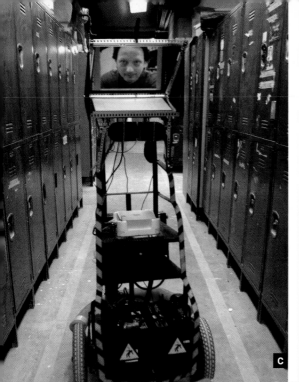

Fig. C: Marque Cornblatt's Sparky 2 projects its user's presence anywhere the robot can find a wi-fi connection. Fig. D: The Bovalve controller combines the concepts of a violin bow and trumpet valve system to play a synthesizer programmed in Pure Data. Nate Bliton developed the Bovalve interface as an independent project in "Designing Experience" at Michigan State University.

joystick to enable the Bovalve to achieve some rather non-trumpet-like effects such as pitch bending and glissandos.

The right hand controls a simulated string player's bow, replicated in the Bovalve by a ribbon suspended from two wheels. A hand-operated actuator slides along the ribbon, allowing the Bovalve player to apply pressure that mimics the use of a bow.

Bliton needed a way to interface between the potentiometers sensing the states of the bow device and trumpet joystick, and Bliton's computer-based synthesizer. His solution? He used the MAKE Controller to interpret position information and feed it to his computer. He also hooked up several LEDs to the controller output ports to indicate when the "valves" were activated.

"I'm still fine-tuning the instrument," Bliton says. "I plan on building a new body out of aluminum and plastic. And as with any new instrument, I'm still learning to play it."

William Gurstelle is the author of five books including *Whoosh Boom Splat — The Garage Warrior's Guide to Projectile Projects*. He doesn't have or want a normal job.

➕ For more projects based on the MAKE Controller, check out the gallery at makingthings.com/projects.

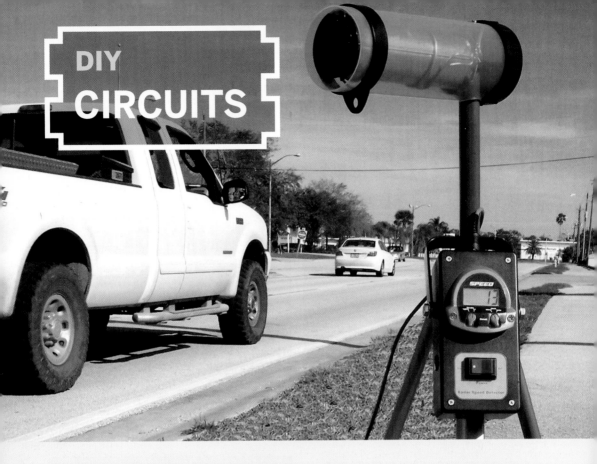

RADAR SPEED DETECTOR

Turn a Hot Wheels toy into a versatile radar gun. By Ken Delahoussaye

I was browsing through a department store one day, in search of a gift for my 8-year-old daughter, when I came across Mattel's Hot Wheels Radar Gun ($30). The box said that this toy could clock the speeds of not only miniature Hot Wheels cars, but also full-sized vehicles.

I figured the toy must have severe limitations, but decided to buy one for my daughter anyway. It turns out that she (we) loved it, and we found that it could accurately measure the speeds of toy cars, cars on the road, even joggers. To my amazement, the detector even measured the speeds of spinning objects like bicycle wheels.

Operating the toy is simple: you aim the gun, squeeze the trigger, and then read the detected speed on the LCD in back. Hold the trigger down for a while and then release, and you'll get the maximum

speed detected during that time. A switch selects either mph or kph readings, and another switch toggles the display units between 1:64 scale (for Hot Wheels) and 1:1 scale actual speeds. Power comes from 4 AAA batteries housed in the handle.

Inside, the Mattel gun uses Doppler radar, just like the expensive speed detection systems used by law enforcement. It transmits a continuous wave at 10.525GHz, then measures the frequency that returns after the wave bounces off a moving target. The main functional difference between the Mattel toy and a $1,000-plus pro model is detection range, which for the toy maxes out at 40 feet. I suspect that this keeps the microwave emissions low enough to guarantee child safety.

Limitations aside, I realized that this so-called toy offers some interesting prospects. The wheels in

Photography by Ken Delahoussaye

my mind began to churn, and I decided to purchase another unit for my own use. I disassembled the gun and decided to repackage it to appear more professional — looks really *are* everything.

I separated the detector component itself (the waveguide antenna) from the display and control panel, then connected the two with a length of instrumentation cable. This configuration lets you position the antenna close to traffic on a tripod, and operate it remotely from a safe distance.

1. Disassemble the toy.

Disassembly is no easy task, as there are 12 screws, each concealed with plastic inserts. I used a drill and ¼" bit to carefully drill out the inserts and gain access to the screws. Use extreme caution with the drill, since some of the screws are located very close to sensitive internal components.

After removing all the screws and opening the case (Figure A), you'll see the long, cylindrical waveguide antenna and a small circuit board that carries the buttons and LCD panel. (The waveguide is a hollow tube that surrounds the microwave antenna and directs and concentrates its signal.)

After recording their locations so you can reattach them correctly later, unsolder all the wires that connect the waveguide, battery compartment, and trigger switch. Be careful when working with the waveguide, which is made of a thin, dielectric material and dents easily. After detaching all the wires, remove the waveguide and display panel, and set them aside.

2. Upgrade the antenna housing.

For the waveguide antenna housing, I chose an "Ice Tube" by Alvin. I've used these 3" diameter acrylic document tubes in previous projects, and I like them. First and foremost, they look cool and futuristic — transparent with various tints. They're also fairly rigid and you can easily cut them with a hacksaw. I used transparent green tubing and cut an 8¾" length.

For a post that holds up the housing and makes it attachable to a tripod, use a ¾" diameter threaded PVC nipple, 8" in length. Drill a ¾" hole in one end of the acrylic tube and insert the nipple through it. Then screw a shrub sprinkler head onto the end of the nipple. Just below the sprinkler head, drill a ¼" exit hole to route the cable through.

Secure the sprinkler head to the wall of the tube opposite the hole, using the screw that came with the sprinkler head plus 2 smaller screws on either side to prevent rotation. For these, drill ³⁄₃₂" holes through the tube and head, and apply two ⅛"×¼" sheet metal screws (Figure B).

To secure the waveguide centered inside the tube, use 12-gauge steel wire, carefully wrapped around the waveguide in a spring-like fashion to form 2 mounts, one at each end (Figure C). Pressure and friction on this wire are what holds the waveguide in the tube housing. There are no screw attachments.

Before inserting the waveguide with mounts into the housing, solder on a new, longer interface cable that will support remote operation. The toy originally used a 4" length of shielded 2-conductor cable, so I figured that the new cable should also be shielded.

In my garage, I found a 20' length of instrumentation cable with shielding. Cut and strip one end of the cable, and solder 2 of the available 4 signal

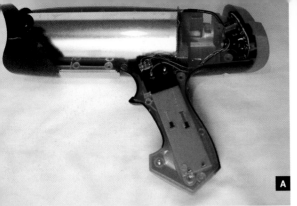

A B

C D

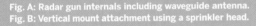

Fig. A: Radar gun internals including waveguide antenna.
Fig. B: Vertical mount attachment using a sprinkler head.
Fig. C: Waveguide with coiled wire mounts.
Fig. D: Waveguide cable wiring to PCB.

wires to the signal contacts on the antenna's round printed circuit board. Also solder the uninsulated drain wire, which connects to the shielding, to the board's ground contact in the middle (Figure D).

Route the opposite end of the cable through the waveguide housing and PVC nipple, and then carefully push the waveguide assembly into the housing (Figure E). To attach the housing to a camera tripod, remove the tripod's head assembly and support tube from the base, then route the antenna cabling through the hole, and slide in the PVC nipple and waveguide housing.

The other end of the cable would attach to the display, but I wanted to make it easily detachable for transport. So I obtained a 5-pin DIN male connector plug and matching jack. Solder the 3 conductors that were connected at the other end to 3 contacts on the DIN plug.

3. Build the display housing.

To house the radar detector's display and controls, originally on the back of the toy gun, I used a 6"×3"×2" project case. First I made a carrying handle out of a bar of 0.1" thick steel (aluminum would have been easier to work with).

Drill ¼" holes in each end of the steel and bend

it into a U shape. Drill corresponding holes in the display housing and attach the handle using hex bolts, nuts, and washers. One of the nice things about the handle is that it also functions as a stand, allowing hands-free viewing when the display unit is on a tabletop or other flat surface.

I needed to make a large hole in the lid of the display housing to fit the LCD display panel. Since I didn't have any large-diameter bits, I used a ¼" bit to make a series of small perforations in the plastic. Punch out the section with diagonal cutters, and file the edges smooth. Then install the LCD panel to the lid by running a couple of ⅛"×¾" sheet metal screws through its 2 original mounting holes.

To replace the functionality of the original momentary trigger switch, I chose a double-pole, double-throw rocker switch. This has the advantage of allowing for automatic, hands-free speed measurements. To install the switch, drill a ¾" hole into the display housing lid, centered below the display, and secure the switch in place with its retaining nut. To reinforce the display and switch, you might also add some glue.

On the top of the display housing, install the female DIN connector receptacle, for plugging in the cable. Drill a ¾" hole into the top of the display

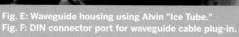

Fig. E: Waveguide housing using Alvin "Ice Tube."
Fig. F: DIN connector port for waveguide cable plug-in.

Fig. G: DIN connector jack mounting, installed in the top of the display housing. Fig. H: Power switch wiring.

housing and attach the jack on the inside using a pair of 3mm×8mm screws and nuts (Figures F and G).

The original radar gun used 4 AAA batteries. This would be fine for our new configuration, but since there is plenty of room inside the display housing, I decided to go with AAs, which have twice the capacity of their AAA cousins.

A quick measurement with my multimeter revealed that the radar system draws about 24mA from the batteries connected in series. This translates to almost 100 hours of continuous use with AA alkalines (assuming 2200mAh each). Install a 4-AA battery holder inside the display housing using double-stick mounting tape.

With all the hardware installed inside the display housing, it's time to solder all the connections. The wires connect to 3 components: power switch, batteries, and the DIN connector for the antenna. For the DIN jack, the toy's original 4" antenna cable had 3 wires, red, white, and a shielding conductor covered with black heatshrink. Solder these 3 wires to the connector contacts matching the other end of the cable, using a multimeter to make sure you get the correspondence right.

The toy's power switch wiring consists of 4 individual wires: 2 brown and 2 blue. Position and solder these to the DPDT switch, following the same pattern as the original manual trigger switch (Figure H).

Finally, the toy's battery wiring consists of 2 individual wires, red and black. Solder these directly to the corresponding contacts on the battery holder (Figure I). That's it for the wiring. After closing up the case and plugging in the antenna cable, you're ready to go (Figure J).

4. Let the fun begin!

I started out by having a family member walk in front of the antenna unit. At a normal pace, the display registered 2 mph, and a more brisk pace yielded a reading of 4–5 mph.

Next, I decided to try measuring the speed of a spinning bicycle wheel. The tripod was handy for this test, since it let me focus the waveguide precisely where I wanted. I placed the bicycle upside down on the floor, supported by its handlebars and seat. Then I turned the pedal to get the rear wheel spinning, and I clocked it at a maximum rate of 15 mph. So far, so good.

Then it was time for the real test, with actual cars on the road. I took the unit outside near the street and set up the tripod. It wasn't long before a vehicle came along, and the readout showed 19 mph. That

Fig. I: Interior of display panel unit with batteries, handle retaining nuts, and DIN connector jack.

Fig. J: Completed display panel unit.

was great — the detector was showing a speed for a passing automobile. But I wondered about its accuracy, so I decided to get into my car and drive down my street past the detector myself. Watching my speedometer, I drove at a steady speed of 21 mph. Heading back to my driveway, I was anxious to see what the device had measured, and to my delight the reading was 21 mph! Mission accomplished.

I've done several projects over the years, and this one has been one of the most satisfying. The Hot Wheels Radar Gun has some great hardware inside, and for my minimal investment, I now have a system that provides a useful function that I'm sure I'll be using and enjoying quite a bit for years to come.

Ken Delahoussaye is a software engineer/consultant in Melbourne, Fla., specializing in embedded firmware and PC applications. He also runs Kadtronix (kadtronix.com), which features automation and control resources.

Empower Your iPod.

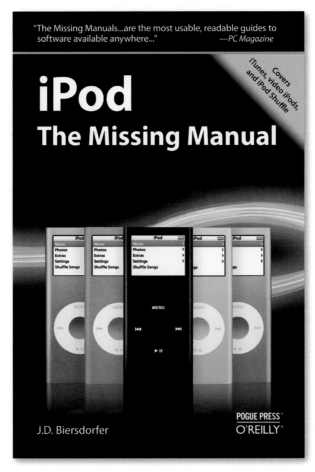

> "The Missing Manuals...are the most usable, readable guides to software available anywhere..." —*PC Magazine*

Covers iTunes, video iPods, and iPod Shuffle

iPod
The Missing Manual

J.D. Biersdorfer

POGUE PRESS™
O'REILLY®

ISBN 0-596-52978-3
256 pages
$19.99 US /$25.99 CAN

With the new iPods, Apple has given us the world's smallest entertainment center. Sleek, powerful and somewhat addictive, these little gems can do far more than play music. To make the most of your iPod's capabilities, pick up a copy of the new *iPod: The Missing Manual*. This new edition thoroughly covers the redesigned iPod Nanos, the video iPod, the tiny Shuffle and the overhauled iTunes 7. Each page sports easy-to-follow color graphics, crystal-clear explanations, and guidance on the most powerful and useful things your iPod can do.

O'REILLY®

Spreading the knowledge of innovators   **www.oreilly.com/store**

Das Bottle

MATERIALS:

2 PLASTIC WATER BOTTLES

SCISSORS

3 RUBBER BANDS

6-INCH RULER

NEEDLENOSE PLIERS

2 PAPER CLIPS

CHOPSTICKS

DRILL HOLE IN BOTTLE CAP.

STRAIGHTEN PAPER CLIP, AND FEED THROUGH BOTTLE CAP THEN THROUGH PROPELLER'S CENTER HOLE. BEND A HOOK INTO SECOND HOLE.

MAKING THE PROPELLER:

CRUSH, THEN CUT A BOTTLE IN HALF.

TRIM AROUND THE BOTTOM.

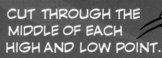

CUT THROUGH THE MIDDLE OF EACH HIGH AND LOW POINT.

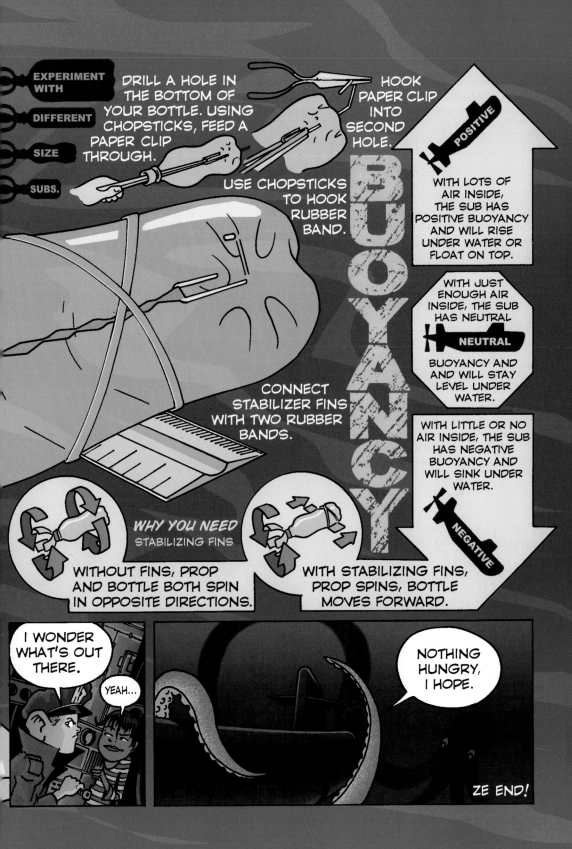

MakeShift

By Lee D. Zlotoff

The Scenario: Thinking you could use a new hobby to get you off the couch, your significant other gifts you with a metal detector for Christmas. After digging up loose change in your backyard and at the beach for a few weekends, you decide it's time for a real prospecting adventure. So, after loading up your SUV with the metal detector, a pick, shovel, pry bar, and enough snacks and water for the day, you both drive 80 miles out into the desert to poke around some abandoned gold mines you've heard about.

Finally reaching the end of the road in the middle of nowhere, you ask your partner to unload the car while you head for the rocks to survey the landscape. But as you climb through some old barbed-wire fencing to look for a trail, your keys — attached to your Swiss Army knife — fall out of your pocket and skitter off across the rocks before they disappear into a deep, 6-inch-wide crevice between two boulders. (Don't you just hate when that happens?) Needless to say, neither your cellphone nor your OnStar system gets reception out here, and the fancy anti-theft option you went for makes hot-wiring your SUV all but impossible.

The Challenge: Without transportation, you're stranded. To avoid the daunting prospect of walking back out to the main road — as well as "never hearing the end of this" from your mate — you're going to need to recover those keys.

The boulders are too massive to be moved in any way and you don't have a direct sight line to your keys. But you are able to ascertain that the depth of the narrow crevice can't be more than 15 feet. It's about noon now, so you've got at least 6–7 hours of daylight to work with before it gets dark. Surely someone with your skills and ingenuity can get those suckers out of there in time to get you home safely, if not still salvage the outing, no? As the wheels start turning, your mate appears and asks, "Is something wrong, honey?"

Here's what you've got: In addition to everything mentioned, there's a basic tool kit in the car: hammer, screwdriver, adjustable wrench, snippers, pliers, etc., as well as 100 feet of nylon rope. Because this is an old mining area, there may also be some small pieces of wood and metal lying around.

Send a detailed description of your MakeShift solution with sketches and/or photos to makeshift@makezine.com by Aug. 17, 2007. If duplicate solutions are submitted, the winner will be determined by the quality of the explanation and presentation. The most plausible and most creative solutions will each win a MAKE sweatshirt. Think positive and include your shirt size and contact information with your solution. Good luck! For readers' solutions to previous MakeShift challenges, visit makezine.com/makeshift.

Lee D. Zlotoff is a writer/producer/director among whose numerous credits is creator of *MacGyver*. He is also president of Custom Image Concepts (customimageconcepts.com).

Electronic Test Equipment

See and understand what's happening inside a circuit.

By Tom Anderson and Wendell Anderson

Take a look at a printed circuit board. You can see components such as resistors and capacitors, but where is the voltage? Where is the signal? How do you tell if the circuit is working correctly? What if you want to change it?

Electronic test equipment lets you probe and "see" the voltages and currents running through electronic circuits. This article covers four basic devices: oscilloscopes, power supplies, function generators, and multimeters. Learning to use these tools — especially the mighty oscilloscope — requires patience, but it's an absolute requirement for building, troubleshooting, and hacking electronic gizmos.

Photography by Sam Murphy

OSCILLOSCOPE TYPES

The oscilloscope ("scope" for short) is an important and expensive piece of test equipment. Its main function is to draw a graph of changing voltages. The time scale goes from left to right on the graph, and the voltage (as measured on the probe tip) goes up and down. The time and voltage scales are adjustable to accommodate slow, fast, small, and large signals. Because audio signals, video signals, and digital logic are all represented by changes in voltage over time, the scope can suss out most interesting circuits.

Analog oscilloscopes, the older and less expensive type, show the signal in real time: if a signal is there, you can put it on the display. You set an analog scope to trigger when specified voltages are detected, or when an external device gives the signal. These scopes work fine for audio signals, but are much harder to use with digital signals, so don't spend your money on an analog scope if you'll mainly be working on microcontrollers or USB or other digital circuits.

A *digital oscilloscope* adds to these capabilities by converting digital signals into a signal "trace," storing it in memory so you can inspect it. You can also export the trace to a computer. Most digital scopes also have an "autoscale" button that does some of the knob twisting for you. For more money, you can look at very long traces, and do frequency analysis (fast Fourier transform) right on the oscilloscope itself. Most important, you can look at one-time events: bursts of data, infrequent signals, or, with a high-end scope, even specific commands or address references.

BUYING AN OSCILLOSCOPE

You'll find hundreds of used analog scopes for $50 and up on eBay. The cheapest ones will need to be fixed, which is very difficult without a working scope. Look for the brands Tektronix, HP, Agilent, Hitachi, Hameg, and Leader. If probes aren't included, find them in a separate auction. A new analog scope will cost $250+, and Amazon carries 4 models.

For digital scopes, a new entry-level model will cost $1,000 and up; you may be able to find one used. High-end digital scopes go for $5,000 and up. Designers of digital scopes do their best to emulate analog "feel," so learning on an analog scope is still relevant, and the entry price is much lower.

Look for a scope with at least 2 channels —

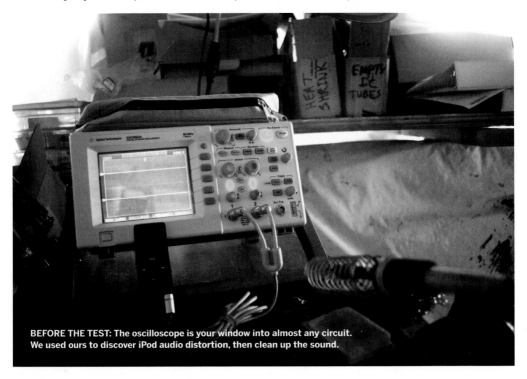

BEFORE THE TEST: The oscilloscope is your window into almost any circuit. We used ours to discover iPod audio distortion, then clean up the sound.

4 is better. Do not buy an *analog storage scope* in order to learn about oscilloscopes, because these are much harder to use. Get as much scope as you can afford; like good books, good tools are cheap if you use them.

POWER SUPPLY

A *power supply* (aka *bench supply*) provides "known-good" power to your circuit. After we figure out how much power our circuit actually needs, we can then decide how to power it for regular use, like with batteries or a wall wart.

Bench supplies let you adjust the output voltage, which is good for initial testing: while slowly increasing the voltage to your circuit's operating value, you can look for burning components, smoke, arcing, explosions, etc. The fanciest supplies also have *current limiting*, which sets the maximum amount of current the circuit is allowed to draw. A circuit that draws way too much current could be cooking itself, or might have a power short to ground — maybe through that screwdriver you left under the PC board. Some power supplies also have meters that show voltage and current.

Many circuits require both positive and negative DC voltages. To test these, you need a *dual-output supply*, which has separate V+, V-, and ground terminals. Expect to spend a minimum of $50 for a good supply, more if you want dual output and meters.

MULTIMETER

A *multimeter* measures the voltages and currents in a circuit, as well as the characteristics of individual components such as resistors and diodes. They all measure resistor values (ohms), but try to find one that can also measure capacitance (microfarads), since it is sometimes difficult to read the values on capacitors.

Use a multimeter to "ohm out" an unknown circuit. This means tracing out the connections on a circuit board to see what is connected to what, so that you can create a schematic drawing that shows all the circuit's connections.

Multimeters are designed to withstand wide ranges of inputs, so you can use one to check for high voltages in unknown circuits. The terms *volt-ohm meter (VOM)*, *multimeter*, *ohmmeter*, and *voltmeter* all refer to pretty much the same thing. Expect to spend a minimum of $10 for a handheld voltmeter,

or $50 for one with capacitance ranges.

FUNCTION GENERATOR

A *function generator* injects a regular signal at a selectable frequency into a circuit. In certain circumstances you can do this with a CD player or other gear, but a function generator is more flexible. Look for one that can generate sine waves (handy for testing audio circuits), square waves (handy for testing filters, amplifiers, and digital circuits), and triangle or sawtooth waves (handy for graphing circuit behavior). More costly units can produce any waveform shape you want; these are called *programmable waveform function generators* or *arbitrary waveform generators*. Expect to spend $75–$500, or more for ones that can generate arbitrary waveforms, or go up to higher frequencies.

HOW TO USE AN OSCILLOSCOPE

The oscilloscope draws a graph of voltage versus time on the display screen (known as a *graticule* to old-timers). By connecting probes to your circuit and generating these scope traces, you can see what's going on.

Don't be intimidated by the large number of knobs and buttons. These controls are grouped into 3 basic function areas: *timebase*, *sensitivity*, and *trigger*. Timebase and sensitivity controls change the graph's appearance, setting the horizontal and vertical axes, respectively. Trigger controls tell the oscilloscope when to start drawing the trace. Here's how to operate each control area.

Timebase — Adjust the Horizontal

Find the biggest knob on the front of the scope, and "ratch" it around clockwise and counterclockwise. This is the *horizontal scale adjustment*, and it determines the time represented by each grid line (or "division") on the display. A typical range is from 100 nanoseconds to 1 second per division. At the short end of this scale, the trace zips left to right so fast on the display that all you see is the line left behind. At the long end, the trace appears as a traveling dot. At any setting, the trace starts at trigger time, which you can think of as time=0.

Sensitivity — Adjust the Vertical

Most scopes have multiple channels, which plot separately to the display. Each channel corresponds to its own electrical input, usually a BNC connector

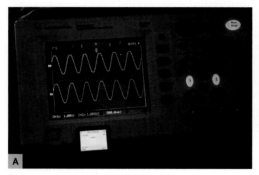

A

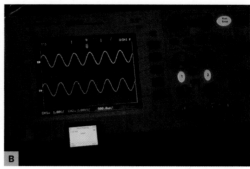

B

on the front panel. A knob for each channel adjusts the *volts/div (volts per division) scale* of the graph, and turning it to "1" sets the horizontal gridlines 1V apart, going up and down. The *baseline voltage* is controlled by another knob, labeled Position or Vertical Position. This lets you move the trace up and down. You can use these knobs to look at more than one trace at a time, adjusting the displayed voltage ranges for each channel to superimpose them against one another.

Most scopes also have an *input coupling switch* that allows you to choose between AC, DC, or GND for each channel's input. GND is for adjusting the oscilloscope; set your ground to 0V. DC mode displays the time-varying signal in absolute voltage, while AC shows the signal centered over the x-axis, adding or subtracting any constant voltage to show how the signal differs from a baseline average.

Trigger — Fire When Ready

Triggering is the hardest part for beginners to learn. One method is to read all about the trigger modes and how they work; the other is to "video-game it" and just keep knob-twiddling until you see the trace. Here are the most common types of trigger conditions, but there are many others:

Normal/single trigger — Triggers whenever the input voltage exceeds the knob-adjustable threshold.
Auto — The scope guesses when to trigger, like the auto setting on a camera.
TV trigger — Special triggering for television signals.
External trigger — Trigger from an electrical signal via a cable plugged into the front panel.
HF reject — Trigger when steadily rising voltage exceeds threshold, but not if it suddenly spikes.
LF reject — Trigger on sudden spikes but not slowly varying voltages.

Slope — Sets whether the trigger detects rising edges or falling edges.

PROBES

Oscilloscope *probes* connect your circuit to the oscilloscope electrical input. Most have a little hook that can grab onto the component's leads on a circuit board. Unscrewing the hook reveals a sharp point that can dig into a circuit board's copper traces directly. Probes also have a dangling alligator clip, which you connect to ground to provide a baseline 0V for your measurements. Poor grounding makes for bad measurements.

The most common scope probes are *10x probes*, which reduce (attenuate) the input signal to one-tenth their original voltage, so a 10V signal will appear on the display as a 1V signal. (You might wonder why these are not called 10% probes.)

TRY IT! TEST AN AUDIO SOURCE

We heard distortion when we first hooked up an iPod mini to our big old stereo. Using a scope, we tracked down the problem. We wired the iPod headphone jack to the scope inputs, and played a test file of a pure 1kHz sine wave. With the iPod volume turned all the way up, we saw that the peaks of the wave were flat rather than round. This distortion is called "clipping." Slight clipping makes sound a bit muddy, and more pronounced clipping makes it fuzzy and clearly distorted. The onset of clipping, where it's slight, is much easier to see on an oscilloscope than it is to hear (Figure A).

Turning the iPod's volume down to 90% eliminated the problem, and generated a nice clean sine wave. From this we learned that you should set your iPod to no more than 90% volume when hooking it up to a stereo (Figure B).

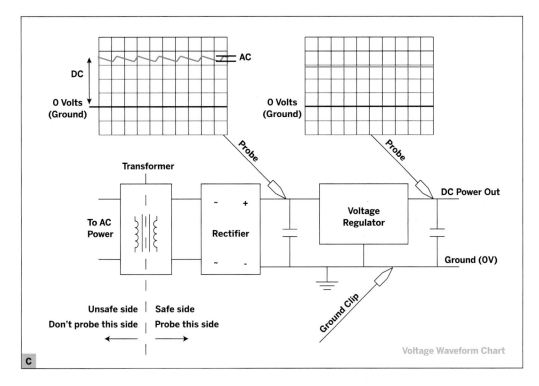

DC

AC

0 Volts
(Ground)

0 Volts
(Ground)

Probe

Probe

Transformer

~   +

Voltage
Regulator

DC Power Out

To AC
Power

Rectifier

~   -

Ground (0V)

Ground Clip

Unsafe side | Safe side

Don't probe this side | Probe this side

Voltage Waveform Chart

C

SCOPE LIMITATIONS

Can an oscilloscope measure any signal? The short answer is no. The longer answer depends on how much money you are willing to spend. An oscilloscope must be able to respond faster than the input signal comes through, and some signals, such as high-bandwidth digital transfers, may change too fast for slower, cheaper scopes to follow. For audio work, almost any scope will work because the frequencies are relatively low, but for video and high-speed digital, expect to spend more.

Measuring voltages below 0.1mV and above 50V will also cost extra, and the ability to trigger on complex data, for example a particular command on a USB connection, costs much more.

BASIC TROUBLESHOOTING STRATEGY

Here's the strategy we generally follow to figure out a circuit, working or not. For this example, we'll analyze an audio amplifier.

First we try to find a schematic; the internet has lots of them. If we can't locate one, we "ohm out" the circuitry while it's powered off, using the multimeter, and gradually draw a schematic that contains all components and connections. This takes practice, patience, and multiple tries.

Then we turn the amp on and check the voltage coming from the power supply, first with the meter, and then with the scope. It's amazing how many

circuits you can fix just by understanding and fixing the power supply. The power supply voltages should show up as an AC waveform at the transformer output. After the rectifier, the voltages should be relatively constant with time, except for some ripple. After the voltage regulator, the voltage waveform should be flat as a pancake (Figure C). Look for noise, or artifacts left over from AC, which can cause hum. If the circuit doesn't have its own power supply, use a bench supply.

Next, we probe with the scope, starting at the inputs and working toward the outputs. Our amp needs an external input, so connect a function generator or other audio source such as an iPod to the input to create a test signal. Test tones for your iPod are available at makezine.com/08/ibump. Start with a very simple signal; for audio input, we use a sine wave. The signal in an amplifier should get larger as you approach the output. If it dies midway through, you've found a problem, such as a dead amplifier stage.

Finally, examine the amp's output. You don't want to see any DC there, or else it might blow out the speaker! If your amplifier has this problem, fix it before it destroys any more innocent speakers.

➕ More resources at makezine.com/10/primer

Brothers Tom and Wendell Anderson started Quaketronics (quaketronics.com) in Tom's garage to share their projects with MAKE readers. The company has since expanded to Wendell's garage. Their kits are available at store.makezine.com.

VOLTAGE, CURRENT, AND RESISTANCE By Joe Grand

Voltage, *current*, and *resistance* are 3 staple quantities that you'll encounter with anything that has electrons running through it. This article explores voltage and current, then resistance, and at the end shows how they all tie together.

CHARGE, CURRENT, VOLTAGE, AND POWER

Electricity starts with *charge*, which means electrons. Charge is measured in coulombs (C), where one coulomb equals $6.25 * 10^{18}$ electrons, a very large number. Electric charge can be static, as held in a capacitor, or it can flow, as through a wire. Flowing charge is *current*, denoted with the symbol I. The rate of flow is measured in amperes, or amps (A), where 1 ampere equals 1 coulomb flowing through a given point on a circuit per second.

Voltage, also known as *potential difference*, is the amount of work (energy) required to move a positive charge upstream in a circuit from a more negative point, which has lower potential, to a more positive point, which has higher potential. Think of it as an electrical pressure, or force. Voltage is represented in equations by V, E, or U, and its unit of measure is the volt (V).

Power is a measurement of the energy that's expended by a flowing charge over some amount of time. One watt of power (W) is equal to the work done in 1 second by 1 volt moving 1 coulomb of charge. You can calculate the power consumed by a circuit with the simple formula:

$$P = V * I \quad \text{where } P = \text{Power (W)}$$
$$V = \text{Voltage (V)}$$
$$I = \text{Current (A)}$$

To differentiate between voltage and current, we refer to voltage as going *between* or *across* 2 points in a circuit, and current as going *through* a device or circuit connection. You say, "The voltage across the resistor is 1.7V," and "The current through the diode is 800mA." When referring to the voltage at a single point in a circuit, it is defined with respect to ground (typically 0V).

DIRECT CURRENT (DC) AND ALTERNATING CURRENT (AC)

Direct current (DC) flows in one direction through a conductor, as either a steady signal or as pulses. The most familiar form of DC supply is a battery. Most electronic circuitry uses DC, except for power supplies or motor circuitry.

Alternating current (AC) flows in both directions, which makes it more complicated to work with than DC. But it travels better, which is why AC is the form of electricity delivered as house current. In the United States and Canada, AC power outlets provide 120V AC at 60Hz (cycles per second). Other parts of the world use different standards.

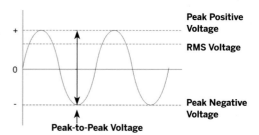

Several terms describe an AC signal:

Peak voltage (V_{PEAK}) — the maximum positive and negative voltages of the AC signal, measured from the center.

Peak-to-peak voltage (V_{PP}) — the total voltage swing from the most positive to the most negative point of the AC signal.

Root-mean-square (RMS) voltage (V_{RMS}) — the most common term used to describe AC voltages, calculated by taking the square root of the average of the integrated square over 1 cycle. This calculation sums both positive and negative voltage, to measure total useful energy.

For a typical sinusoidal AC signal, you can use the following 3 formulas:

$$V_{PP} = 2 * V_{PEAK}$$
$$V_{PEAK} = 1.414 * V_{RMS}$$
$$V_{RMS} = 0.707 * V_{PEAK}$$

RESISTANCE

Electricity can be described using the analogy of water: coulombs are gallons, amperes are gallons per second, voltage is psi of water pressure, and wattage is still wattage, because both electrons and water can do work. Following this analogy, resistance is like pipe width. The narrower a pipe, the more it restricts water flow (current), which means more resistance. If a pipe is large (low resistance), then water (current) can flow through more easily. At any resistance, higher pressure (voltage) will force more current through. Current that's prevented from flowing by high resistance leaks out and is dissipated as heat, which results in a voltage difference across the resisting conductor.

Some resistance exists in any electrical device. *Resistors* are components manufactured to have a known, fixed amount of resistance, which are used to reduce the current flowing through a point in a circuit (Figure A). Resistors are not polarized, meaning that they can be inserted in either direction with no change in electrical function. Inside, the resistance comes from carbon, metal film, or wound wire. A *potentiometer*, or "pot," is simply a variable resistor, and you typically change its resistance by turning a knob.

Resistors are defined by 3 values:
Resistance (in ohms, designated with the omega symbol, Ω)
Maximum heat dissipation (in watts, W)
Manufacturing tolerance, accuracy of labeled resistance (%)

The resistance and tolerance values are indicated on each resistor by a standard code of colored bands (Figure B).

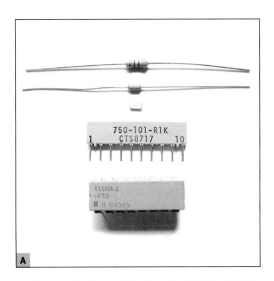

A

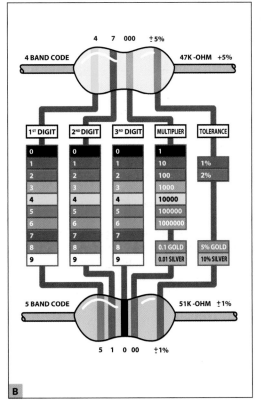

B

If you're not good with remembering the color table, a useful mnemonic is "Better Be Right Or Your Great Big Venture Goes West," which corresponds to black, brown, red, orange, yellow, green, blue, violet, gray, and white, the order of increasing value. There are other mnemonics less suitable for printing.

Most general-purpose, carbon-filled resistors allow a ±5% tolerance on the resistor value. Metal film resistors are more precise and usually have a ±1% or ±2% tolerance, which suits them for applications such as amplifiers, power supplies, and sensitive analog circuitry.

Resistors typically come in wattage values of $\frac{1}{16}$W, $\frac{1}{8}$W, $\frac{1}{4}$W, $\frac{1}{2}$W, and 1W, which refer to how much power they can safely dissipate. Typical resistors are $\frac{1}{4}$W and $\frac{1}{2}$W, and big, wire-wound resistors can handle more. To calculate the required wattage value for your resistor application, use one of the following 2 equations:

$$P = V \times I$$
$$P = I^2 \times R$$ where P = Power (W)
$\qquad\qquad\qquad V$ = Voltage across the resistor (V)
$\qquad\qquad\qquad I$ = Current flowing through the resistor (A)
$\qquad\qquad\qquad R$ = Resistance value (Ω)

MULTIPLE RESISTANCES IN SERIES AND IN PARALLEL

When resistors are used *in series* in a circuit, their resistance values are additive, meaning that you add the values of the resistors to obtain the total resistance. For example, if R1 is 220Ω and R2 is 470Ω, the overall resistance will be 690Ω (Figure C).

In contrast to resistors connected in series, multiple resistances wired *in parallel* decrease overall resistance, because they provide alternative pathways for current flow. The overall resistance given by resistors in parallel (Figure D) can be calculated by:

$$1 / R_{TOTAL} = (1 / R1) + (1 / R2) + (1 / R3) + ...$$

For 2 resistors, this formula becomes:
$$R_{TOTAL} = (R1 \times R2) / (R1 + R2)$$

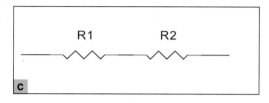

C

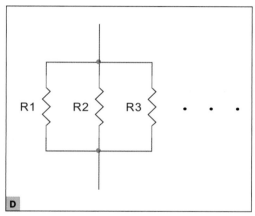

D

OHM'S LAW

Ohm's law is a basic formula that ties it all together, summarizing the relationship between voltage, current, and resistance in an ideal conductor. It states:

$$V = I * R$$ where V = Voltage (V)
$\qquad\qquad\quad I$ = Current (A)
$\qquad\qquad\quad R$ = Resistance (Ω)

Understanding these basic theories and running the numbers from your multimeter through these equations will let you confidently explore unknown pieces of hardware, or even create new inventions.

Origami Flying Disc
By Cy Tymony

Understand Bernoulli's principle of flowing fluids and gases with a paper flyer.

You will need: One 8½"×11" piece of paper, scissors, transparent tape

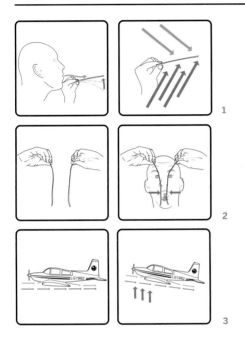

1. Demonstrate Bernoulli's principle.

Cut a paper strip ½"×4". Hold the paper strip just below your lips, and blow above the strip (Figure 1). The paper will rise!

This occurs because of Bernoulli's principle — fast-moving air has lower pressure than non-moving air. The still air below the strip has higher pressure than the moving air above, so it pushes the strip upward.

Try another quick test: cut 2 paper strips ½"×4". Hold them about 2" apart and blow air between them (Figure 2). You expect them to blow apart, but they actually come together. Bernoulli's principle is working here because the faster-moving air between the strips has lower pressure than the air outside them, which therefore pushes them together.

The top of an airplane wing curves upward and has a longer surface than the bottom. When the plane moves, the air moving along the top must travel farther and faster than the air moving past the straight bottom. The faster-moving top air has less pressure than the bottom air. This provides lift (Figure 3).

2. Make an origami flying disc.

Now use your newfound Bernoulli principle knowledge to make a flying disc using only paper and tape.

Cut eight 2"×2" square pieces of paper (Figure 4). Fold the top right corner of one square down to the lower left corner (Figure 5). Then fold the top left corner down to the lower left corner to create a small triangular pocket (Figure 6). Repeat these two folds with the remaining 7 squares (Figure 7).

Insert one paper figure into the left pocket of another (Figure 8). Insert the figures into each other until they form an 8-sided disc. Hold the disc firmly together while you apply tape as required (Figure 9).

3. Fly it.

Turn the disc over and toss it like a Frisbee. You'll see that it glides a bit, but then drops rapidly. The reason is that the top and bottom surfaces are straight.

Now bend down the outer edges to form a curved lip (Figure 10). This should produce the Bernoulli effect.

Throw the flyer with a quick snap of your wrist, and it should stay aloft for a much greater distance.

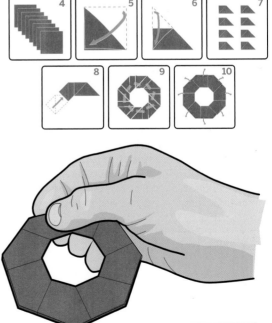

Illustrations by Tim Lillis

Get started in electronics, eliminate red-eye the old-school way, and touch up your walls with the screw of a lid.

TOOLBOX

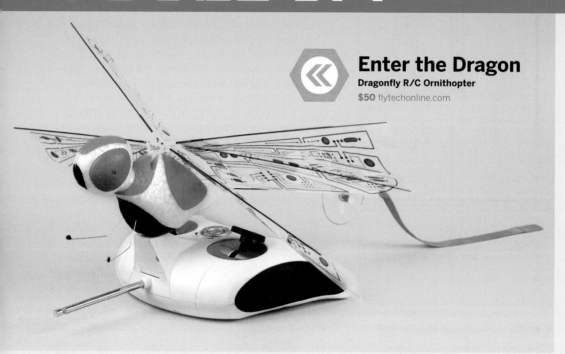

Enter the Dragon
Dragonfly R/C Ornithopter
$50 flytechonline.com

WowWee continues its string of high-tech toy hits with the Dragonfly, a genuine R/C ornithopter. WowWee likes to find talented designers and entice them to Hong Kong to make killer-cool toys. First they wooed Mark Tilden, who created the Robosapien. Then they turned to high school student Sean Frawley, who they found selling orni-thopter kits online. Now responsible for the FlyTech line at WowWee, Frawley's first creation is the Dragonfly.

The Dragonfly is extremely light. The 16" wings use carbon fiber struts, and the styrofoam body weighs just 0.8oz. The flier is powered by a lithium polymer battery recharged via AA batteries in the controller; you get 10 minutes on a charge. The digital radio controller has two channels, for beginner and expert modes; expert allows for tighter turns. The Dragonfly can travel up to 18 mph within a 50' R/C range.

As with Robosapien, one of the coolest things to do with a Dragonfly is to hack it! Early hacks include removing the styrofoam to reduce weight, replacing the LED eyes from blue (which draw ~3V) to red (~1.6V) to save battery life, and tracking down the Chinese battery manufacturer to discover a longer-life version (which may be flyable in a denuded Dragonfly). For other hacks, go to robocommunity.com.

—*Gareth Branwyn*

Four Eyes and Ears
SoundVision Safety Glasses
$20 fullpro.com

Here in the MAKE Lab it's a good idea to wear your ear and eye protection, but all that gear on your melon makes you feel like a deep-sea diver. The SoundVision kit brings you out of the depths and back to your workshop, where you belong. Just slap on your favorite pair of earmuffs and attach the Velcro-strapped safety glasses to avoid crushing the temples of typical glasses into your head. They're comfortable and good at deflecting flying pieces of metal and wood away from your eyes; we checked. The glasses come in clear, smoke, and amber if you're after that stylish shooting-range look.
—Jake McKenzie

Sappy Cards
$10/6 sappycards.com

Dear Toolbox,
A necessity I recommend including in your column: Sappycards ... the right tool for dealing with holidays, awkward crushes, and general communication of love for the socially inept. My next-door neighbor, Tim Furstnau, makes these sarcastic, hilarious cards using designs inspired by security envelope patterns and old book covers. Tim is a linguist and all-around amazing guy with a blunt and refreshing sense of humor that comes across in his cards: "Something always prevents me ... from killing you." You need them in your tool-box, too. *—Andrea Steves*

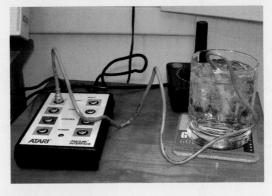

AtariLab
Prices vary ebay.com

When I'm explaining microcontrollers to someone who's around my age, I usually start out with a reference to 8-bit home computers: "Take something like an Atari 800, make it really small, and that's kind of what a microcontroller kit is." I always worried about whether that comparison was close to the mark, but now that I've gotten my hands on an AtariLab kit, I doubt no more.

In the mid-1980s, the AtariLab debuted and included a breakout board that connects to the joystick port, a cartridge, and a collection of sensors. I picked mine up on eBay, and when I connected the sensors, plugged in the cartridge, and turned on my Atari 130XE, it worked perfectly. I was up and measuring temperatures in seconds. (If only all my computers would boot up that quickly.)

The AtariLab kit invites hacking, too. Because it uses the joystick port, it's easy to write programs that read the sensor: LOGO and BASIC source code is included in the lab manual!
—Brian Jepson

TOOLBOX MAKE LOOKS AT KITS

Editors Phillip Torrone and Arwen O'Reilly talk about kits from makezine.com/blog.

LED Button Pad

$20 sparkfun.com

Inspired by Monome's 40k device (*see page 22*), this pad of 16 button-switches shines with tricolor LEDs (sold separately). The pad can be broken into four 4-button pads, too.

No-Friction Bicycle Dynamo

$18 freelights.co.uk/kit.html

Avoid that annoying whir you hear when using a friction bike light. Make an electromagnetic induction generator from parts instead.

Cellular Automata Video Synthesizer Kit

$50 makezine.com/go/cellular

With the completed kit you'll be able to "uncover endless visual and sound patterns on any TV." We like their fun retro enclosure as well — make your own!

Mousey the Junkbot

$20 solarbotics.com

Featured in MAKE, Vol. 02, and on *The Colbert Report*, this famous light-seeking lemming is now available as a parts kit from our friends at Solarbotics. (Mouse shell, battery, and wire not included.)

Kindest Cut of All

$6 colorcutter.com

Florida-based inventor Perry Kaye of Gizmo Enterprises came to Maker Faire and handed out samples of one of his inventions. In Hollywood-speak, think Magic Marker meets X-Acto knife. He put a blade inside a felt tip pen and called it the ColorCutter. Now you can draw shapes and cut them out at the same time, a boon for free-form prototyping.

The ColorCutter seems to work like magic, and it's easy to fool your friends before you explain how it works. With the right kind of pressure on its felt tip, the ColorCutter's blade will cut the paper as you draw. What's cool, however, is a kind of safety feature, which Kaye was anxious to demonstrate. Put the tip of the pen on your finger and the blade doesn't come out and cut you. Nonetheless, a CYA safety warning on the pen says: "Never Touch Tip," which just makes you want to do it. The ColorCutter can be used to cut shapes from paper, plastic sheets, fabrics, and mylars.

—Dale Dougherty

Nail Polish Meets House Paint

$15/3-pack qwikie.com

Sometimes a tool comes along that's so ridiculously simple, but makes a huge impact on how you go about your business. Qwikie Paint Pots are such a wonder-widget. Basically, they're like large nail polish jars that you use to store house paint in for hassle-free touch-ups. When you're finished painting a room, you put some of the leftover paint in one of the Qwikie pots. They have a brush built into their screw-on tops. When a scuff or a chip on a wall calls for a touch-up, it takes seconds to grab the jar and fix the boo-boo.

This is a huge improvement over having to head to the basement, find the right can, pry it open, peel off the "skin," mix it up, find a brush, use it, clean it, put the lid back on the can, etc. You can store all of the colors to your entire house in a shoe box in the hall closet! Maybe a little pricey at $5 each, but still worth it. *—Gareth Branwyn*

Rice Cooker

$30 amazon.com/kitchen

The image of geeks surviving on colas and cold pizza is not completely myth. When you're really cranking on a cool project, who can spare the processor cycles to cook? Well, the genius of this classic rice maker is a thermostatic switch, set to cut the power once all the liquid has boiled. Measure ingredients, push the lever, and get right back to hacking — with no risk of burning the house down.

Asian households have used them for decades to make foolproof rice — fancier models even click over to keep-warm mode when done. But the undocumented tip is this: you can also dump in veggies, spices, meat, and so on, to balance your carbs with other nutrients. And for breakfast, try it with oatmeal. It's cheap, tasty home cooking that runs as a background process.

—Ross Orr

KIDS' CORNER

I got a thing called a **Book Wizard** today. It looks a bit like an oversized LED book light, but it's actually a device that you clip onto your book that has the time, date, timer, and a few more groovy gadgets. A really cool thing that it does is you can type in what page you are on and it will remember that. makezine.com/go/bookwiz

—McKinley Rodriguez, age 10

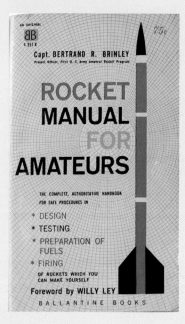

« Blast Off

Rocket Manual for Amateurs by Capt. Bertrand R. Brinley

Price varies, Ballantine Books, 1960

Bertrand Brinley was the author of the famous Mad Scientists' Club books for teens, stories of unflappable high schoolers who built all manner of fantastic stuff. Brinley also wrote an impressively complete guide to designing, building, fueling (!), and launching your own full-sized rockets. His *Rocket Manual* contains everything from nozzle designs to bunker construction plans to extremely detailed notes on propellants — just skimming it motivated me to make sure the zinc and the sulfur in my house were as far apart as possible. It's also full of complex ballistics calculations, engineering formulae, and logarithm on top of logarithm.

In short, it's actual rocket science — but it never once occurred to Brinley that his readers would be incapable of rocket science, or for that matter any complex and difficult thing whatsoever, as long as the desire was there. After your incredulity at his confidence in you dies away, you'll begin to feel that regardless of what your own apparently lofty goals are — from starting your own magazine to building a kit car to knocking over the Bank of Scotland — it would really be a shame to let Brinley down by not even trying.

—*John Krewson*

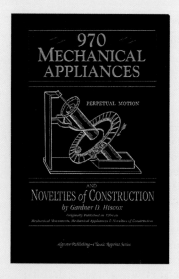

« Tinkerer's Delight

970 Mechanical Appliances and Novelties of Construction by Gardner D. Hiscox

$14, Algrove Publishing

Algrove Publishing offers a reprint series of classic books on engineering and technology. Originally published in 1904, this quality paperback is a great way to explore the historic roots of today's machines and mechanisms. Each entry is illustrated with clear, crisp line drawings from patents and from mechanics' magazines, and describes each device's construction and use. Topics include power, steam, explosives, marine vessels, railway devices, gears, timekeepers, mining, factories, textiles, and construction. Don't dismiss this as just a museum of "obsolete" engineering, for the perceptive reader will find many mechanisms still used today, and others that might deserve a revival in your next basement project.

Of special interest is a concluding chapter of 56 perpetual motion devices. Hiscox includes these, even though he "has not the slightest desire to encourage the hopeless pursuit of perpetual motion," in hopes they might lead "those who still believe in reaching this *ignis fatuus* to bend their energies in causes more worthy of their zeal." Anyone who enjoys making things will surely derive inspiration from this book.

—*Donald Simanek*

《 Game Time

How Computer Games Help Children Learn by David Williamson Shaffer
$27, Palgrave Macmillan

As a game designer and a parent, I like the author's argument that kids learn the most by participating in reality-based games. Once challenged, they do the hard work of researching the scenario and looking at the issues from different angles. Computer games can train kids to think flexibly and for themselves, Shaffer shows, while schools emphasize memorization and compartmentalize disciplines that are — in an innovator's mind — intertwined.

But you must choose games carefully. In *SimCity*, players make decisions dictatorially, which isn't realistic. *Urban Science*, in contrast, is what Shaffer calls "epistemic" because it conveys the experiences actual urban planners go through as they negotiate with the Chamber of Commerce, preservation society, and other interests. Unlike a "god game," *Urban Science* reflects the real world. This is good, but Shaffer's analysis lacks the "how" of building open-ended simulations. Constructing each option takes work, and including them all is impossible.

In the end, Shaffer asks whether our society will make the changes needed to teach children innovation, or continue along the worn path reinforced by No Child Left Behind. Testing will always prove the value of testing. —*Matt Coohill*

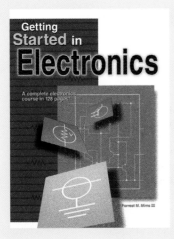

《 Oldie but Goodie

Getting Started in Electronics by Forrest M. Mims III
$20 forrestmims.com

Like many geeks my age, I got my start in electronics with a Science Fair Electronic Project Kit. The projects from that kit taught me the basics, before I strayed from the hobby and followed other pursuits. When I decided to get back into electronics, I had a lot of catching up to do. I found countless recommendations for *Getting Started in Electronics* and picked up a copy of this slim book. It's designed to look like it was written on a lab notebook, and perhaps it was. It's got everything I needed to know: coverage of essential components such as resistors, capacitors, transistors, all the way up to integrated circuits.

It goes way beyond explaining how to use these components and gets into how they actually work, and how to assemble projects (breadboarding, wire wrapping, and soldering). The last part of the book shows you 100 electronic circuits you can build using everything you've learned. What I found amazing about this book was how much information was packed into a small space. The free-form text design and hand-drawn figures work well together. When I first flipped through the book, it looked crowded to me, but once I started reading, it was a constant payoff in knowledge and insight. If only I'd read this 25 years ago. —*Brian Jepson*

❮❮ Designer Flash Drives

One of our staff members (and we're not naming names) is the type of person who thinks nothing bad could ever happen to her. And while we appreciate the optimism, it sometimes leads to bad habits — like not backing up one's hard drive. A few months ago, this staffer's 7-month-old Mac-Book committed suicide and took all her documents with it. Lesson learned. Now this little robot guy from Mimoco keeps extra copies of her work, and also serves as one of our many adorable office mascots.

Rhymezone.com

When brainpower's running low and we're desperate for a clever headline, our last resort is always poetry. This online rhyming dictionary with a thesaurus attached has helped us out in plenty of pinches. An awful headline on fortunate mistakes became "It's a Blunderful Life." A bland title on homeless parades became "The Fragrance of Vagrants." The solutions aren't always genius, but at 3 a.m. RhymeZone always seems to understand.

❮❮ Twirling Fork

When you're as lazy as our founders, the simple task of twirling your own pasta seems like a tremendous chore — which is exactly why there's a spinning fork from stupidiotic.com in our kitchen cabinet. At the flip of a switch, the fork spins your spaghetti into perfect, bite-size pasta bundles, meaning there's no risk of spraining your finger with all that needless twirling action. Best of all, the utensil is dishwasher safe, so you'll never have to handwash it.

Print Magazine

We can't rely on scantily clad celebs on our covers to help sell magazines (the bare-chested Einstein on our Swimsuit Edition being the only exception), so we generally have to get a little more creative with the design. Fortunately, thumbing through *Print* is always inspiring. The mag is jam-packed with the most genius design work around, and if you've got even an ounce of creativity running through your system, skimming the pages will kick-start your neurons and get your brain purring again.

❮❮ Dell Projector

Like many companies, we use a projector — although not for anything as productive as, say, showing Power-Point slides. Truth be told, we bought our Dell 3400MP so we could watch NCAA basketball games at something like real-life size for at least one-third of what we would have spent on a clunky flat-screen. (Give us a break, more than half of us went to Duke.) We can even use the projector for actual work-related stuff, like screening *Psycho*, which we wrote about in our Winter 2006 issue, thankyouverymuch.

❮❮ Hulk Hands

Nothing makes hanging out at the office more fun than throwing on a pair of these ridiculous Hulk Hands. Equipped with tons of great sound effects, the giant foam fists are particularly useful for banging on the table at editorial meetings, Khrushchev-style, while grumbling things like, "Editor mad!"

In 2001, two recent Duke grads set out to create an intellectual magazine for the anti-intellectual. The result: *Mental Floss*, the perfect antidote to dull conversations. mentalfloss.com

Rules of the Game

Quasicade
$599 quasimoto.com

I learned all the essential life skills growing up in the local arcade, from martial arts to shooting zombies. One of my goals in life, ever since my first birthday party at Chuck E. Cheese's, was to have a game room in my house, complete with an arcade cabinet. Sadly my dreams were dashed when I realized how much it costs to buy each game.

Then came the Quasicade from Quasimoto: an arcade cabinet made to order. It came in three boxes, ready to assemble, and in under an hour I was looking at a thing of beauty. It connects to any home game station or computer, and all it needs after that is a monitor or TV. As soon as I got those hooked up, I was having a blast. It stands in a corner of the MAKE Lab, and is a hit with everyone who plays it. The games are limited only by how many you and your friends have lying around at home.

—Matthew A. Dalton

The Power of Convenience

USBCell Rechargeable AA Batteries
$19 for 2 usbcell.com

Here's an idea I wish was mine. These USBCell batteries are NiMH rechargeables with a built-in USB plug so you can plug in and charge up using any one of the world's billion or so USB ports. I'm fairly certain I am not the only person on Earth who loathes all those giant power bricks out there, so do your part and get rid of one.

—Jake McKenzie

Arcades Aplenty

GameRoom Magazine **$36**/year
gameroommagazine.com

From pinball cabinets to countertop touchscreens, rebuilt hits from years long past to future machines in the works, *GameRoom* is a fantastic magazine for getting all the information you want on arcades and home game rooms. Plus, with all the sample articles they have online, you can see how much you like the magazine before you make that purchase, which is really pretty sweet of them.
—*Matthew A. Dalton*

Music nerd/sound engineer **Andrea Steves** showcases her green home restoration at mongodeco.com.

Donald Simanek's Museum of Unworkable Devices can be found at www.lhup.edu/~dsimanek.

Gareth Branwyn is the cyborg-in-chief of streettech.com.

John Krewson is an editor at *The Onion* and dreams of rockets.

Jake McKenzie is one half of the MAKE engineering intern duo.

Matthew Dalton comprises the other half of the MAKE engineering intern duo.

Matt Coohill is an 11-year game design veteran.

Ross Orr keeps the analog alive in Ann Arbor, Mich.

Have you used something worth keeping in your toolbox? Let us know at toolbox@makezine.com.

Editor's Note: It was with great pleasure that we received the following two letters from young readers. We hope you enjoy them as well.

■ How MAKE Magazine Changed My Life

Vincent Brubaker, age 13
1/4/07

Personal Statement

I got MAKE magazine on Christmas Eve in 2005. I received it as a 1-year subscription from a family friend, Richard Lawler. At first, the magazine seemed too good to be true. It was an amazing DIY (Do-It-Yourself) magazine with more than an imaginable amount of projects, tips, tricks, ideas, how-tos, places to get things, places to go, and things to do, electronic as well as non-electronic. After I got the magazine, I started to make all sorts of fun electronic circuits and projects, and hack away at toys until they do things that they wouldn't normally do.

The Maker Faire is something that MAKE magazine has started to do annually; it's a big festival where people come to show off their stuff that they think would be worth sharing to people who read MAKE magazine or to people who are just interested. I went to the first annual Maker Faire with Richard, my mom, and another friend, Joe, and had a blast! I had fun doing all sorts of activities, looking at displays, and taking tours of mini-exhibits.

I got to the jackpot. It was a huge building full of electronics that you were allowed to scavenge. To my dismay that I couldn't have done this earlier, the Maker Faire was closing down — as well as ending their two-day run. I made off with all sorts of goodies! I got two 8,200µF 80V capacitors, a programmable bicycle wheel message displayer DIY kit, and to top it all off, I got two 32kµF 50V screw-in-lead type capacitors that were each the size of two small cans stacked on top of each other.

The Maker Faire is important to me because I am able to meet hundreds of people who like to do the things that I like to do. It is also important to me because I am able to learn a lot about the things that I already do, and see how other people do the things that I like to do — their techniques, methods, etc. — so I can get ideas from them.

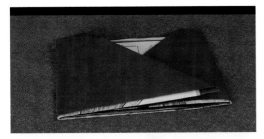

On the Contributors page in Volume 09, we regrettably misspelled the name of star MAKE intern Jake McKenzie.

The article "TV Set Salvage" (*Volume 09, page 138*) had an error in a photo caption. Figure D shows a deflection yoke, not a flyback transformer. Figure B shows a flyback transformer, with the red, second anode lead still attached.

Although we contacted Tom Valone to verify the content of our profile of him (*Volume 09, page 50, "Patently Curious"*), some errors remained. Valone obtained his master's degree in physics in 1984, taught in the physics department of Erie Community College, began working at the Patent Office in 1996, and resides in Maryland. We apologize for the inaccuracies.

A photograph in "Primer: Carbon Fiber" (*Volume 09, page 164*) incorrectly shows the mandrel being rounded on a router sitting upside down. This can be dangerous — to rout safely, you need the support that a router table provides.

MakeShift author Lee Zlotoff writes:

A special Honorable Mention Award to the third-graders in the After School Science Club at Chapman Hill Elementary School in Salem, Ore. Teacher Maureen Foelkl and her students Savannah Brown, Marisa Chen, Emily Farnell, and Cory Francis took on this challenge as a science project and submitted an elaborate and impressive set of materials that contained an imaginative solution. MacGyver would be proud of you all!

Check out makezine.com/makeshift to read more of Zlotoff's comments and all the winning entries.

■ Entry for the MakeShift Contest, Volume 08 (stranded on a deserted island with no drinking water)

I have fallen from our boat into the deep blue sea. I grasped my gear, and I felt like a drowning swimmer. A sneaker wave pulled me up in the air. I fell unconscious on an island. It must have been at least five hours until I woke up. A crab pinched my nose. Ouch! If I caught that crab I would be eating him for dinner. My mind was filled with the thought of that great breakfast I had this morning. I wondered if they serve French toast here for breakfast.

I noticed the only things I saved were my matches, my Swiss Army knife, a piece of sailcloth, and the clothes on my back. It was time to explore the island. I imagined there would be a hotel with a pool and a water slide. The kid meal would be a buffet of chocolate bars, assorted candy, sugar cubes, triple layer cake, and of course a chocolate fountain.

It was time to find this hotel. I started walking toward the tip of the island. It was kind of small. I'd never been in a place where everything had an ocean view. I stopped. I cried. I was so disappointed. There was no hotel. Where were the people? Where were the restaurants? Where's my mommy? I was really thirsty. There wasn't any fresh water around.

Our group, the After School Science third-graders, came up with a brilliant idea on how to make fresh water. Look at our poster and follow the instructions. If you can follow all the directions you can make some fresh water yourself.

So go out there into the world and find yourself a stranded island. Good luck!

—*Savannah Brown, Marisa Chen,
Emily Farnell, and Cory Francis*

In the Beginning Was the CRT

By George Dyson

Once upon a time, there was no distinction between memory and display.

Introduced in 1897, the cathode-ray tube brought us oscilloscopes, television, radar, computer terminals, the electron microscope, and, 110 years later, YouTube. But the hum of flyback transformers, by which so much code was written, is at an end. As the last generation of warmblooded monitors vacates our desks, let us remember that the cathode-ray tube's contribution to digital computing began as internal memory, not external display.

Conventional CRTs display the state of a temporary memory buffer whose contents are produced by the central processing unit (CPU). Once upon a time, however, cathode-ray tubes *were* the core memory, and *they* stored the instructions that drove the operations of the CPU. This was one of those sudden adaptations of pre-existing features for unintended purposes by which evolution leaps ahead.

By 1953 there were 53 kilobytes of random-access memory in the entire world, with 5kB in the original IAS machine.

In 1945, when John von Neumann began to orchestrate the electronic computer project at the Institute for Advanced Study in Princeton, N.J. (*see MAKE, Volume 06, page 190*), there was no high-speed random-access memory available off the shelf. Vladimir Zworykin and Jan Rajchman at RCA agreed to supply a plug-and-play digital memory tube, christened the Selectron, an electronically switched array of 4,096 separate targets storing one binary digit each. After two years, there were still no Selectrons in existence. "They were doing things inside that vacuum that hadn't been done before," says Willis Ware, one of the original engineers. A 256-bit Selectron was eventually produced in limited quantities, but too late to compete with magnetic-core memory, and it barely achieved the historical footnote it deserves as a missing link between the vacuum tube and the integrated circuit.

The IAS team decided to improvise a random-access memory from commercially available parts. Existing high-speed storage was based on acoustic delay lines, developed for identifying moving targets by distinguishing radar signals that had shifted from one moment to the next. A series of electrical pulses, about a microsecond apart, were converted to a train of sound waves circulating in a long tube of mercury equipped with crystal transducers at both ends. About 1,000 binary digits could be stored in the millisecond it took to travel the length of a five-foot "tank." The delay line spawned the first generation of serial-access stored-program electronic computers (*see MAKE, Volume 08, page 178*), although "its programming," as Alan Turing's supervisor Max Newman noted, "was like catching mice just as they were entering a hole in the wall."

If you wanted one particular bit, you had to wait a full millisecond and catch it as it went by. How could you read or write any bit at any time? Researchers at MIT's Radiation Laboratory had noted that digital information could be stored as charged spots on the face of ordinary cathode-ray tubes, as long as the pattern was regenerated a few times a second by a trace from an electron beam. The spots become positively charged (i.e., deficient in electrons) as a result of secondary electron emission by the phosphor, and the state of an individual spot could be distinguished by briefly "interrogating" that location and noting the character of a faint secondary current, of less than a millivolt, induced in a wire screen positioned close to the outside face of the tube. "Thus the phosphor containing the various charge distributions is capacitively coupled to the wire screen," the IAS team reported, "and it is then possible by focusing the beam at a given point to produce a signal on the wire screen."

Frederick C. Williams, after working on pulse-coded IFF (Identification Friend or Foe) radar systems in England and the United States, had developed a serial-access cathode-ray memory tube in 1946. In June 1948, he constructed a small computer at Manchester University, under the direction of Max Newman and assisted by Alan Turing, that demonstrated CRT-based storage and a rudimentary stored program. The so-called "Williams tube" memory was highly sensitive to electromagnetic disturbances, and was plagued by the presence of an electric traction line that produced stray magnetic fields.

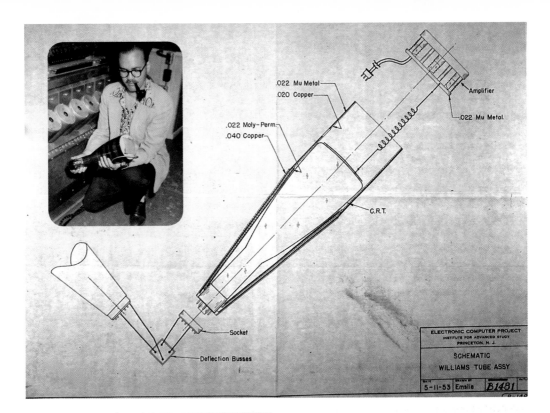

.022 Mu Metal
.020 Copper

.022 Moly-Perm.
.040 Copper

Amplifier

.022 Mu Metal

C.R.T.

Socket

Deflection Busses

ELECTRONIC COMPUTER PROJECT
INSTITUTE FOR ADVANCED STUDY
PRINCETON, N. J.

SCHEMATIC

WILLIAMS TUBE ASSY

5-11-53 | Emslie | B1481

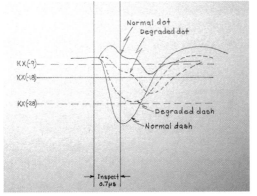

Normal dot
Degraded dot

KX(-9)
KX(-18)

KX(-28)

Degraded dash
Normal dash

Inspect
0.7μs

ABOVE: Lead engineer James Pomerene, holding one of the IAS Electronic Computer Project's electrostatic storage tubes, with cutaway view. Each tube has a capacity of 1,024 bits, with 20 tubes (visible behind Pomerene's shoulder) arrayed like half of a V-40 engine on each side of the machine.

LEFT: When a designated address in the 32×32 array was "interrogated" by a pulse from the electron beam, a faint secondary current induced in the wire screen attached to the face of the CRT indicated either a dot (0) or a dash (1), with less than one microsecond to discriminate between the two.

Random access might be possible if suitable timing and control circuits for the electron beam deflection voltages could be engineered. Chief engineer Julian Bigelow paid a visit to Williams in Manchester in July 1948, and the IAS team, led by James Pomerene, soon developed switching circuits that could read or write to any location at any time, appropriating a few microseconds before resuming the normal scanning cycles where they left off. The resulting memory organ was, in effect, an electronically switched 32×32 array of capacitors, but was, as Bigelow noted, "one of mankind's most sensitive detectors of electro-magnetic environmental disturbances." Errors were introduced by fields of as little as .005 gauss, or $1/40$ the strength of the Earth's magnetic field.

The IAS group settled upon standard 5-inch 5CPIA oscilloscope tubes, available in quantity, although in 1953 it was reported that "there had not been more than ten flaw-free tubes discovered in the testing of over 1,000 tubes in this laboratory during the past three years." The manufacturers allowed the IAS to scan their inventory for unblemished specimens and ship the others back. The ability to distinguish a dot (0) from a dash (1) depended on the secondary emission characteristics of the phosphor coating, and the slightest imperfection would cause the memory to fail. The faint signal was amplified 30,000 times before being passed to a discriminator that made a decision as to whether the waveform represented a 0 or a 1.

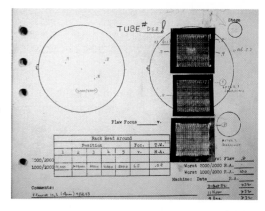

Pomerene achieved a 34-hour error-free test of a two-stage memory on July 28 and 29, 1949, and the final race to build a working 40-stage memory began. All 40 memory tubes had to work perfectly at the same time. Data was processed by operating on all the digits of a 40-bit word at once, with each bit assigned the same position in a different Williams tube, an addressing scheme analogous to handing out similar room numbers in a 40-floor hotel. This made the computer 40 times as fast as a serial processor, but, in the opinion of numerous skeptics, unlikely to work without one thing or another always going wrong. "The rig can be viewed as a big tube test rack," Bigelow observed.

A 41st monitor stage could be switched over to mirror any of the 40 memory stages, allowing the operator to inspect the contents of the memory to see how a computation was progressing — or why it had come to a halt. This was later augmented by a separate 7-inch cathode-ray tube serving as a 7,000-points-per-second graphical display.

Each individual memory tube had its own logbook recording its health history and any idiosyncrasies that arose along the way. The memory constantly had to be "brought back into focus" and the resulting difficulty in distinguishing memory problems from coding problems drove many early programmers near-insane.

"The presence of this leprous element in the machine [means] that everyone who sits down to do a problem must be aware of it and be prepared to be just a little cagey, depending on the problem," complained one early user, "for the blips fade in a fraction of a second and if the problem requires that you re-use a number before the blip is regenerated, you get the wrong answer. It is as if a desk calculator would fail any time the 7th, 8th, and 9th places in a 15-digit number happened to be a three-digit prime … it just isn't decent for the operator to have to worry about how the machine is built."

Nonetheless, the Williams tube changed the world. Both the Williams group in Manchester and the von Neumann group in Princeton agreed to preclude patent disputes by placing the invention in the public domain. More than a dozen first-generation copies of the IAS machine were built, and the second generation included the IBM 701.

Thanks to a small gang of nonconformists, programmers were given a true random-access memory, where any storage location could be addressed at

TOP IMAGE: Logbook page for an individual memory tube, 1952–1953. The "screenshots" pasted into the log are direct photographic images captured for diagnostic purposes. By 1953 there were 53 kilobytes of random-access memory in the entire world, with 5kB in the original IAS machine.

BOTTOM: Machine log, 11:45 p.m., 9 September 1954: "Raster suddenly expanded" and entire memory has turned to "garbage." With a diagnosis of "suspect deflection circuitry," the engineers (who were running a numerical evolution experiment for Nils Barricelli) shut down at 12:16 a.m.

any time. The digital universe as we know it came into existence when the address matrix was freed from the physical restrictions that serial access had imposed. Only then could code start freely moving around. All hell broke loose as a result.

Someday, and it may be soon, the flickering light of the last cathode-ray tube monitor will fade to black, never to return. For a few seconds, an electrostatic charge will linger on its surface, a ghostly memory from a time when cathode-ray tubes ruled the world.

George Dyson, a kayak designer and historian of technology, is the author of *Baidarka*, *Project Orion*, and *Darwin Among the Machines*.

Antique Computers Run the World
By Tom Owad

Photograph by Dennis van Weeren

■ I was standing in the control room

of a power station, staring at the frozen screen of the computer that controls the turbine system. Reset it, and I risked corrupting a hard drive containing the only working copy of the long-obsolete turbine control software. We had called the manufacturer for support and received the simple reply, "We no longer support that model." No software, no parts, not even any advice was available for a 15-year-old computer responsible for keeping a 500-megawatt power station operational. I hit reset and held my breath, as I watched the system reboot. With the system back online, I made an image of the hard drive and tried to work out the system's idiosyncrasies so we could build a replacement for it.

The situation is not uncommon. I have another customer that's currently stockpiling old Macs so they can continue to run a sophisticated custom application they began developing in 1984.

Hobbyists and historians face similar difficulties. Whereas the technology in an antique piece of machinery is visibly apparent, a dead computer tells one very little. The contents of the firmware and the logic of the custom ICs slip away, leaving a dead black box.

I had these problems in mind recently, while devising a system to automate some 1960s manufacturing equipment for a customer. We've gone to great lengths to use open source hardware and to ensure that we have all the schematics we'll need to reproduce the circuit boards if they're discontinued. One board uses a PIC microcontroller, of which we're about to order a lifetime supply.

Open source hardware designs are becoming common. Open Circuits (opencircuits.com) provides downloads and links to schematics and printed circuit board layouts, and program files for numerous open source hardware projects. Opencores.org has a large repository of Verilog and VHDL chip logic designs (known as "cores"). These cores can be loaded onto FPGAs (field-programmable gate arrays) as they stand, or can be integrated into one's own design.

Some of these projects are a boon to classic computer enthusiasts who dread the day when their machine boots its last. Syntiac's FPGA 64 is an entire Commodore 64, written in VHDL (a design language for FPGAs), that can be loaded onto an FPGA or the C-One microcomputer. For Apple

▲ Dennis van Weeren's tiny Minimig emulates the venerable Amiga 500 personal computer.

users, Alex Freed took the 6502 in Syntiac's design and combined it with work of his own to create the FPGApple, a single-chip clone of the Apple II.

Dennis van Weeren is nearing completion of his Amiga 500 re-implementation, the Minimig. It boasts an exact re-creation of the Amiga's custom chipset, and it currently runs on the Xilinx Spartan-3 starter kit, along with two printed circuit boards containing the M68000 processor and an MMC memory card slot that supports the FAT16 file system and can decode ADF disk images on the fly.

Code is also available for the Z80, 8088, and many other processors and components. The FPGA in the C-One can be reprogrammed to change the system from a Commodore 64 to an Amstrad CPC. A similar FPGA-based computer, the 1chipMSX, is also under development, with plans to support the Amiga, Amstrad CPC, Apple II, Atari ST, and Commodore 64 and VIC-20 computers.

With the logic of these systems being preserved in open-source VHDL and Verilog files, it may be that if you want a program that will still function 50 years from now, your best bet is to write it on a classic 8-bit microcomputer.

Tom Owad is the owner of Schnitz Technology, a Macintosh consultancy in York, Pa. He spends his days tinkering and learning, and is the owner and webmaster of applefritter.com.

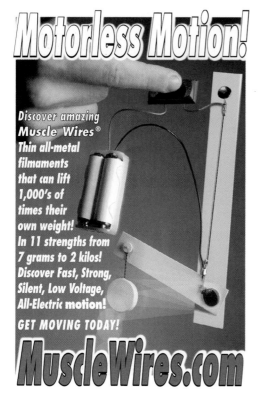

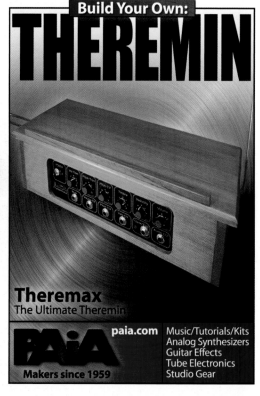

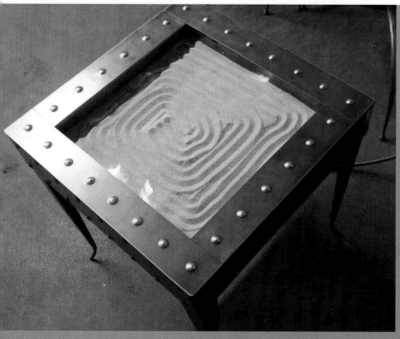

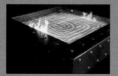

MAKER'S CALENDAR

Compiled by William Gurstelle

Our favorite events from around the world.

Jan	Feb	Mar
Apr	May	Jun
July	Aug	Sept
Oct	Nov	Dec

»» JUNE

»» Power of DC
June 3, Mason-Dixon Dragway, Hagerstown, Md.
Fast, powerful, and silent electric vehicles race for ecologically sound glory. nedra.com

»» Minnesota Inventors Congress June 8–11, Redwood Falls, Minn.
A conference for makers of all sorts. Features renewable energy innovations, electric cars, zubbles (colored bubbles), and a chance to meet Jay Leno. inventhelper.org

»» National Concrete Canoe Competition
June 14–16, Seattle
Since 1971, engineering students compete to see who can build the best concrete canoe. asce.org/inside/nccc2007

»» Montréal International Fireworks Competition
June 20–July 28
Montréal, Québec
The best fireworks-making teams compete in what may be the most prestigious pyrotechnics contest in the world. internationaldesfeux loto-quebec.com/en

»» Paris Air Show
June 22–24, Paris, France
The Salon International de l'Aéronautique et de l'Espace is a huge trade fair for the aerospace industry but welcomes visitors to catch the latest in aviation. paris-air-show.com

»» Field of Flight Air Show & Balloon Festival
June 29–July 4, Battle Creek, Mich.
Large, action-packed air show featuring the Blue Angels, the Shockwave Jet Truck, and lots and lots of balloonists. bcballoons.com

»» JULY

»» RoboCup 2007
July 1–10, Atlanta, Ga.
RoboCup is a combination of robotic events (soccer, robot rescue, and robots at home) that promotes robotics and related fields. robocup-us.org

»» LDRS 26
July 12–17,
Jean Dry Lake, Nev.
One of the premier amateur rocketry events in the world, featuring hundreds of high-power rocket launches. ldrs26.org
»» Related story on page 48

»» Da Vinci Days
July 20–22, Corvallis, Ore.
Da Vinci Days, a community festival celebrating the crossroads of art, science, and technology, is a tradition in this town. davinci-days.org

»» RoboGames
June 15–17,
Fort Mason Festival Pavilion, San Francisco
A festival of robotics, with more than 70 events, including robotic combat, soccer, sumo, and more. robolympics.net

Photography by Nate Smith and Dave Patterson (The Crucible), and Debbie Dannenfelzer (Bonneville Speed Week)

>>**Fire Arts Festival**
July 11–14, The Crucible,
Oakland, Calif.
A unique festival for
fire makers from around
the country. Event
includes fire-related
classes, installations,
performances, youth
classes, and an open
house. thecrucible.org
>> Related story on page 42

EDITOR'S CHOICE
>>**EAA AirVenture**
July 23–29, Oshkosh, Wis.
More than 750,000
people attend AirVenture,
the annual convention
of the Experimental
Aircraft Association.
Thousands of aircraft-
related events.
airventure.org

>> AUGUST

>>**DEFCON**
August 3–5,
Riviera Hotel, Las Vegas
The venerable DEFCON
remains one of the
largest and deepest
computer hacker con-
ventions in the world.
defcon.org

EDITOR'S CHOICE
<<**Bonneville Speed
Week** August 11–17,
Bonneville Salt Flats, Utah
The fastest racers on
Earth streak across the
salt flats attempting
new speed records.
scta-bni.org

>> SEPTEMBER

<<**International
Bognor Birdman**
September 1–2,
Bognor Regis, U.K.
A competition for
human-powered flying
machines held on
England's south coast.
A prize of £25,000 is
offered for the farthest
flight over 100 meters.
www.birdman.org.uk

Important: All times, dates,
locations, and events are
subject to change. Verify all
information before making
plans to attend.

*Know an event that should be
included? Send it to events@
makezine.com. Sorry, it is
not possible to list all submitted events in the magazine,
but they will be listed online.*

*If you attend one of these
events, please tell us about it at
forums.makezine.com.*

See your future at ⊙

Sometimes, it takes more money to buy it than to make it from the money itself.

 $.06
U.S. Mint Spec
Copper-Plated
Zinc Alloy

 $7.02
Dime Store
Brass-Plated
Pot Metal

Photograph by Tom Parker

Nerds in Space
By Bre Pettis

Makezine.com's Weekend Projects shoots for the stratosphere.

FIVE SNAPS UP: Four cameras point outward to capture a 180° panoramic image from space. Using the MAKE Controller, they take a picture every 7 seconds. A fifth, hacked CVS camera on the bottom gets the first 20 minutes of flight on video.

Every week I publish a Weekend Projects video that teaches you how to make something. Sometimes it's a straightforward project like how to make a workbench, and sometimes it's an ambitious collaborative project like my recent near-space balloon project code-named AHAB — A High Altitude Balloon.

For this podcast I needed help, and my friends at the Public N3rd Area (PNA) workshop in South Seattle stepped up to hatch a plan. The idea was to send a weather balloon and its payload up higher than airplanes can fly, to take pictures where the sky is black and you can see the curvature of the Earth. Near-space weather balloon flights have been done before, but we had some ideas to make this flight special.

To take a super-wide panorama, we packed in 4 cameras and a MAKE Controller to fire them off every 7 seconds. We added a hacked CVS video camera and pointed it downward to get the first 20 minutes of the ascent on video.

We built two different tracking systems. The first was a standard beacon that used a ham radio, GPS receiver, and TinyTrak position encoder to broadcast location data to local repeaters and then onto the internet. Our backup system was an inexpensive cellphone loaded with Mologogo software,

which beacons GPS data over the cell network. Some clever programming made all of this available for the world to watch live on Google Earth.

Using a free wiki at balloon.pbwiki.com made collaborating on the project easy. Each person was responsible for researching and building a component for the launch, and we aggregated all the intelligence we had onto the wiki. Without the wiki, this project would have been a nightmare to organize. And it now stands as a document for anyone else to follow in our footsteps and recreate the project.

Our first attempt to fly the balloon, on March 3, was averted due to weather. We got out to Coulee City and found snow and dense fog. The FAA requires at least 50% clear skies, and we couldn't even see 40 feet. We settled on a tethered test and sent the payload up 150 feet. All systems checked out, and we learned a lot about how to make it better. When we got back, we made a number of improvements.

On April 7, we returned to eastern Washington, and a month made all the difference. Temperatures were in the 70s and skies were clear; we were all set for takeoff. We were in constant contact with the FAA and local airports to let them know what we were up to, so they could keep airplanes out of the neighborhood of our launch.

Photography by Bre Pettis

HELIUM-FILLED (Clockwise from left): Latex gloves kept the balloon pristine during inflation. Folks around the world watched the flight in Google Earth in real time. Bre Pettis with the big mill at Seattle's Public N3rd Area, where friends meet weekly to hack hardware and software.

We did final tests, and with all systems go, we launched the balloon. As the balloon goes up, the air gets thinner and the balloon expands until it gets so big that it bursts and shreds. Just in case it stabilized in the jet stream, we had a backup system, which would count down 2½ hours and then cut the cord, sending the balloon up faster and dropping the payload to Earth with its bright orange parachute.

We watched the APRS position data flow in and quickly noticed a few things. It wasn't rising as fast as we had planned, and it was going higher than we had hoped! We were astounded as it flew to 80,000, then 90,000, then 100,000, and then 109,000 feet. At this point the cut-down device switched on and severed the line to the balloon, and the payload dropped. Unfortunately, our excitement turned to sadness when the tracking devices stopped transmitting at 60,000 feet. The -40°C (-40°F) temperatures and slower-than-planned ascent had depleted our battery and we lost track of the payload!

We had a last known position and estimated velocity, and with the help of a friendly local pilot, we flew around searching in her Cessna 182. Despite covering 100 square miles, we couldn't find it. Within a day, high-altitude ballooning friends from around the globe asked for our wind data and

crunched the numbers. We now think the payload, with all those great pictures in it, is located just be found. It's got my number on it and the SD cards have been known to survive going through the laundry, so I'm pretty sure the pictures will be there when it turns up.

To learn more, read about it on the blog at makezine.com/blog and watch the Weekend Projects podcast at makezine.com/podcast.

UPDATE: Our original guess was a bust, but thanks to a crack team of math ninjas, we have new points to play with; feel free to help search!

Point A: 47.7157313, -119.8058445
Point B: 47.7164263, -119.8002014
Point C: 47.7166049, -119.7976196
Point D: 47.72626377, -119.79456

Photos of search locations:
📷 flickr.com/photos/bre/451959324/
flickr.com/photos/bre/451959290/

Bre Pettis produces MAKE's Weekend Project Podcast. Tune in every Friday afternoon and learn about a project you can make in a weekend at makezine.com/podcast.

My 5-Foot Radio-Controlled Submarine

By Michael Wernecke

HOMEBREW

During early mornings at a swimming pool, I ran my 22-inch radio-controlled submarine. I liked these peaceful moments as much as I liked building things. This gave birth to a new dream: I decided to build a submarine from scratch.

I liked the look of the Russian Alfa-class sub and sent for some plans. There aren't many photographs of the Alfa submarine, and the drawings and the DOD photos conflicted. I had to do some guessing, but I did have some general ideas. The sub would have a container of compressed air onboard to blow water out of a ballast tank in order to submerge. I would find some bulkhead seals to pass control linkages through. I would make the propeller, and I would build the antenna array ... thus, I proceeded with the build.

Each day I had questions, and a rising financial expenditure, but I didn't want what I had built so far to go to waste, so I kept moving ahead. I made the hull prototype and cast the hull halves in fiberglass and epoxy resin. To connect the two halves accurately and easily, I created a ridge along the inner lip of the upper and lower hull halves. I completed the prototype, sprayed it with primer, and, like fine Tiffany silver, carefully engraved it with hatch, torpedo tube, and vent locations. I made tools I didn't have. I made templates and found a great carbide-tipped tool to scribe the various locations.

I thought I was fighting a major engineering debacle when the time came to mount the watertight cylinder to the lower hull half. I made a bracket-and-clip apparatus that held the cylinder and hull in place at the same time. I cast the ballast weight from lead shot and then glued it to the inside bottom of the sub's lower hull. I spent many hours balancing the sub in a flotation tank, front to back and side to side, practicing surfacing and submerging the ship. I thought I would never get it right.

After adding closed-cell foam for flotation, I finally achieved proper trim. Splendid! I took the ship to the lake and ran it. The ship worked, and I was pleased. It has become a dream-come-true and a source of great pride and satisfaction for me.

Michael Wernecke presented at the Maker Faire in San Mateo, Calif., in May 2007. He creates Alfa hull kits and welcomes questions about his sub. ocean_tech04@yahoo.com

Photograph courtesy of Michael Wernecke